The Macat Library
世界思想宝库钥匙丛书

解析世界环境与发展委员会
《布伦特兰报告：我们共同的未来》

AN ANALYSIS OF
WORLD COMMISSION ON ENVIRONMENT AND DEVELOPMENT'S
THE BRUNDTLAND REPORT: OUR COMMON FUTURE

Ksenia Gerasimova ◎ 著
李莹莹 ◎ 译

目 录

引 言 …………………………………………………… 1
   格罗·布伦特兰其人 2
   《我们共同的未来》的主要内容 3
   《我们共同的未来》的学术价值 4

第一部分：学术渊源 ………………………………… 7
   1. 作者生平与历史背景 8
   2. 学术背景 12
   3. 主导命题 16
   4. 作者贡献 20

第二部分：学术思想 ………………………………… 23
   5. 思想主脉 24
   6. 思想支脉 28
   7. 历史成就 32
   8. 著作地位 36

第三部分：学术影响 ………………………………… 41
   9. 最初反响 42
   10. 后续争议 47
   11. 当代印迹 51
   12. 未来展望 56

术语表 …………………………………………………… 60
人名表 …………………………………………………… 65

CONTENTS

WAYS IN TO THE TEXT	71
Who Is Gro Brundtland?	72
What Does *Our Common Future* Say?	73
Why Does *Our Common Future* Matter?	75
SECTION 1: INFLUENCES	79
Module 1: The Author and the Historical Context	80
Module 2: Academic Context	85
Module 3: The Problem	89
Module 4: The Author's Contribution	95
SECTION 2: IDEAS	99
Module 5: Main Ideas	100
Module 6: Secondary Ideas	105
Module 7: Achievement	110
Module 8: Place in the Author's Work	115
SECTION 3: IMPACT	121
Module 9: The First Responses	122
Module 10: The Evolving Debate	127
Module 11: Impact and Influence Today	132
Module 12: Where Next?	137
Glossary of Terms	141
People Mentioned in the Text	146
Works Cited	149

引言

要 点

- 格罗·布伦特兰出生于 1939 年，是挪威第一位女首相，是联合国*（简写为 UN，促进政府间合作的国际组织）的杰出人物。1983 年，她被任命为联合国世界环境与发展委员会*（布伦特兰委员会）的主席。
- 作为一名训练有素的医生和经验丰富的政治家，布伦特兰认识到环境挑战和人类发展*（旨在为人们提供更多生活选择的政治和社会进程）之间的根本联系，提出了极富远见的方案来管理环境和社会资源。

- 《我们共同的未来》（1987）也被称为《布伦特兰报告》，是可持续发展*（社会在发展的同时保护自然资源的手段）的主要参考文献之一；它提出的"可持续性*"是在学术界和决策圈中被引用次数最多的定义。

格罗·布伦特兰其人

格罗·哈莱姆·布伦特兰 1939 年出生于挪威首都奥斯陆。她的父亲古德蒙·哈莱姆和布伦特兰一样，是一位医生和杰出的政治家。当谈到她父亲对她的职业选择的影响时，她说："我的父亲投票支持工党（挪威工党*：一个社会民主党派），我为此感到自豪，因为这是正确的选择，即使他并没有从中得到什么好处。"[1] 布伦特兰在美国哈佛大学公共卫生学院学医时，接触到了"国际主义的种子"[2]。（这里的"国际主义"是指赞成不同国家和人民之间合作的政治立场。）

布伦特兰毕业后从医，但在 1974 年她搁置了自己的医疗事业，转而担任挪威环境大臣。1981 年，她成为挪威首相，是第一位担任首相职位的女性。她认识到人类健康与环境状况之间的联系。[3] 她是唯一担任过环境大臣职务的首相，这样的背景独一无二。当时的联合国秘书长哈维尔·佩雷斯·德奎利亚尔* 任命她为世界环境与发展委员会主席，领导制定关于环境与发展的"政治行动呼吁"。[4]

布伦特兰召集了一群杰出的委员，包括学者和政界人士，共同提出了关于人类发展的远见卓识，并将观点汇编成报告，题为《我们共同的未来》。根据牛津大学出版社的说法，这份报告是"十年来关于世界未来的最重要的文件"。[5]

《我们共同的未来》的主要内容

《我们共同的未来》是布伦特兰委员会* 受委托制定"全球的变革议程"[6] 而撰写的一份报告，该委员会是联合国召集的一个机构，旨在促进各国之间的合作，以实现可持续发展。它被视为是对 20 世纪 80 年代全球共同体面临的许多严重问题的回应。报告认识到有必要制定管理自然资源的长期战略。布伦特兰委员会的委员们一致认为，在通过国际合作来刺激经济和社会发展的同时，必须要保护环境。综合考虑，这些目标被概括为"可持续发展"，在报告中被定义为人类"既满足当前的需要，又不危及后代满足其自身需要"的能力。[7] 从这份报告的标题中可以看出，我们认识到了对未来所担负的责任。

《我们共同的未来》已被翻译成 20 多种语言，并连同配套的教育视频资料首先分发给政治家们。[8] 该报告呼吁人类要改变其行为方式，包括个人层面和全球层面，以扭转濒临危机的环境和社会趋势。1987 年，牛津大学出版社出版了该报告。这是第一部以联合国

报告形式出版发行的作品。⁹ 至 1989 年，该书已售出 50 万册。¹⁰

《我们共同的未来》将"可持续性"这一术语引入国际政治的日常用语。在学术界，关于可持续发展的讨论催生了一门新的学科，即环境经济学*（研究受可持续性问题影响的经济问题，以及旨在保护环境的政策所带来的经济后果），并受到了其他学科的关注，如人文地理学（研究人类与环境之间的相互作用）、国际关系学（研究国家之间以及国家与国际机构之间的关系）以及社会人类学（研究人类如何建立和维持社会制度，如法律和亲属关系）。《我们共同的未来》还对学术研究和政策制定都有参考价值。

毫不夸张地说，《我们共同的未来》非常重要，影响力极大。报告中提出的观点已被普遍接受，包括联合国大会*（代表世界各国的联合国机构）和世界卫生组织*（也被称为 WHO，负责全球公共卫生的机构）在内的一些联合国机构都来征求格罗·布伦特兰的专业建议。时任联合国秘书长潘基文*任命布伦特兰为联合国秘书长的三位环境气候变化*特使*之一。（这里"特使"指由秘书长选定的，在特定事务上担任代表和顾问的人。）

《我们共同的未来》的学术价值

《我们共同的未来》被公认为是可持续发展这一概念的主要出处。它引入了术语"可持续性"和当代发展政策的概念基石——经济增长、社会发展（诸如生活水平，医疗保健以及法院等机构的改善）和自然保护这三大支柱。虽然布伦特兰和她的委员会成员们不是第一批使用这个术语的人，但他们首次提高了该词在国际上的曝光率，让其成为一个使用率颇高的概念。

报告提出了与平衡人类需求和负责任地管理地球自然资源相关

的普遍问题。文中提出必须要停止过度使用自然资源，这一深刻的见解启发了许多读者。[11] 报告的作者们是国际政策领域中第一批将环境保护与经济发展联系起来的人。他们把满足人类需求作为发展的主要目标，[12] 指出了环境破坏所造成的严重负面影响，同时说明了这些影响如何威胁到当今和未来的人类福祉。

布伦特兰和其委员会的委员们在发表该报告时，列举并分析了人类社会面临的主要挑战；其中包括贫穷、有限自然资源的退化和气候变化。他们还讨论了全球层面和国家层面上可行的解决方案，如加强国际合作、提高政治家和公众对环境的认识。作者们认为报告中所讨论的挑战是全球性的，改变国际政策和提高对这些挑战的认识是可持续性发展的关键所在。

在建立了可持续发展的主要概念框架之后，委员们讨论了未来趋势，并且预言人类将于 2000 年在全球范围内实现可持续性发展。现在看来，当时的预言是过于乐观了。国际社会计划于 2030 年实现可持续发展目标（SDGs）*，尽管这比布伦特兰最初的计划晚了整整 30 年。

布伦特兰的委员们还明确了可能会参与这场变革的机构和人员。他们预测各国政府会抵制他们所提出的激进的变革方案。因此，政策变革的主要推动力有望来自民间团体*——由普通人而非专业政治家组成的政治团体——他们将与公众和决策者共同努力。

---

1. 格罗·哈莱姆·布伦特兰：《女首相：权力与政治生活》，纽约：法勒、施特劳斯

和吉鲁出版社，2005年，第19页。
2. 联合国：《格罗·哈莱姆·布伦特兰博士传记》，日内瓦：联合国，2014年，第1页。
3. 世界卫生组织："格罗·哈莱姆·布伦特兰博士、总干事"，日内瓦：世界卫生组织，1998年，登录日期2016年3月9日，http://www.who.int/dg/brundtland/bruntland/en/。
4. 世界环境与发展委员会：《我们共同的未来》，牛津：牛津大学出版社，1989年，第x页。
5. 世界环境与发展委员会：《我们共同的未来》，出版说明、封面。
6. 世界环境与发展委员会：《我们共同的未来》，第viii页。
7. 世界环境与发展委员会：《我们共同的未来》，第8页。
8. 普拉塔普·查泰吉和马塞厄斯·芬格：《地球经纪人：权力、政治和世界发展》，伦敦：劳特利奇出版社，2013年，第84页。
9. 世界环境与发展委员会："世界环境与发展委员会报告：我们共同的未来"，1987年8月4日，登录日期2016年2月1日，http://www.un-documents.net/wced-ocf.htm。
10. 琳达·斯塔克：《希望之兆：致力于我们共同的未来》，牛津：牛津大学出版社，1990年，第3页。
11. 韦因·佩雷拉和杰里米·西布鲁克：《问地球：印度的农场、森林和幸存者》，弗吉尼亚州斯特灵：地球瞭望出版社，1990年，第62页。
12. 世界环境与发展委员会：《我们共同的未来》，第44页。

第一部分：学术渊源

1 作者生平与历史背景

要点

- 《我们共同的未来》(《布伦特兰报告》)提出了可持续发展的经典定义,沿用至今。
- 该报告引发了国际政治思想的重大转变,促使人们认识到以负责的态度管理自然资源,以造福子孙后代的迫切需要。
- 报告中提到的主要挑战,如气候变化,如今仍具有现实意义。

为何要读这部著作?

格罗·布伦特兰的《我们共同的未来》(1987)是提出"可持续发展"这一概念的重要文本:人类社会应开展各种经济和体制改革,确保人们有更多的生活选择,"既可以满足人们当前的需要,又不危及后代满足其自身需要"[1]。该报告的出版标志着历史上一个重要时期的开始,因为这是高层政治家们首次认识到管理世界自然资源的"老一套"方法最终会导致全球环境的崩溃。作者呼吁在全球层面上重新全面思考生产和分配问题。

该报告首次全面概述了世界人口快速增长、粮食短缺*(食物缺乏)、城市人口增加、环境退化和气候变化等重大问题。截至1989年3月,已有22个国家政府制定了实现可持续性的计划,这表明他们接受了该报告以及可持续发展的新理念。[2] 该报告继续影响着如今的政策制定过程。

> "我决定接受这一挑战,接受面对未来、保卫后代人利益的挑战。因为很明显,我们需要有进行变革的使命感。"
>
> —— 格罗·布伦特兰:《我们共同的未来》主席前言

作者生平

格罗·布伦特兰1939年出生于挪威首都奥斯陆。她的父亲是一名医生,也是挪威杰出的政治家。受父亲的影响,布伦特兰也成为了一名医生。就读于哈佛大学公共卫生学院期间,她提升了自己对于健康和人类发展的认识。她自小便参与政治活动,八岁那年,父亲就让她加入了挪威左翼劳工运动。

布伦特兰在完成哈佛大学研究生学业后,于1965年返回挪威,加入挪威卫生部,担任医疗顾问。1974年,她接受了环境大臣的提名,于1981年成为挪威最年轻的首相,也是挪威第一位女性首相并连任两届(1986—1989和1990—1996)。担任过环境大臣和首相的独特经历,使得布伦特兰成为领导世界环境与发展委员会的不二人选。这个机构是由联合国设立的,意图在资源有限的世界中解决发展这一紧迫问题。

布伦特兰组建了一支国际领导团队,由来自不同国家的21位知名学者、政治家和官员组成,共同撰写《我们共同的未来》。布伦特兰将其委员会的工作描述为基于"友好和开放交流的精神",体现持"不同观点和视野、不同价值观和信仰、不同经历和见解"的人们的学习过程。尽管存在种种差异,但报告的作者们仍达成了共识:可持续发展迫在眉睫。[3]

创作背景

格罗·布伦特兰继承父业，成为挪威著名的政治家，同时也是一位执业医生。在被提名牵头撰写《我们共同的未来》之前，她曾参与联合国其他委员会的工作，处理裁军（停止使用战争武器）、贫困和环境问题。作为一名经验丰富、国际公认的政治家，她熟知当时存在的主要问题。

《我们共同的未来》可以说是时代的产物。它表明我们急需在政治、经济和环境领域的"引人关注的现实"中制定可持续发展的战略任务。[4] 当时冷战*（1947—1991）——美国及其盟国与现在已经解体的苏联*及其盟国之间关系严重紧张的时期——仍在持续。国际社会面临着发展中国家大量公共债务（即国家债务）激增、富裕国家向贫穷国家提供的援助减少，以及贫困加剧等诸多问题。[5] 报告还以非洲的饥荒、1984年的印度博帕尔灾难*（历史上最严重的工业灾难）和1986年的俄罗斯切尔诺贝利事故*（迄今为止最严重的核电站灾难）为例，告诫我们必须改变态度和政策，否则人类的未来将会受到威胁。[6]

报告中提到的许多挑战今天依然严峻，如气候变化、粮食短缺和环境退化。这也说明了为什么格罗·布伦特兰作为联合国时任秘书长潘基文的一位特使，仍然积极参与国际政治，就气候变化问题开展工作。

1. 世界环境与发展委员会:《我们共同的未来》,牛津:牛津大学出版社,1989年,第8页。
2. 琳达·斯塔克:《希望之兆:致力于我们共同的未来》,牛津:牛津大学出版社,1990年,第3页。
3. 世界环境与发展委员会:《我们的共同未来》,第xiii页。
4. 世界环境与发展委员会:《我们的共同未来》,第xix页。
5. 世界环境与发展委员会:《我们的共同未来》,第x页。
6. 世界环境与发展委员会:《我们的共同未来》,第3页。

2 学术背景

要点

- 《我们共同的未来》提出了在自然资源减少的情况下,人类社会如何繁荣发展的重要问题。
- 布伦特兰委员会的委员不同意人口增长会带来灾难性后果这一假设,他们对此持乐观态度,认为可持续性(我们必须生活在有限的自然资源范围内)这一新观念可以解决这些问题。
- 报告展现了作者及其同时代人对人类更美好未来的希冀和理想。

著作语境

格罗·布伦特兰报告《我们共同的未来》(1987)对可持续发展概念的演变做出重大贡献,这一点有目共睹。可持续发展是指,在提高生活水平的同时,确保后代不会在环境退化的世界中失去资源。该报告承认"发展"和"环境"是密不可分的,因为显而易见,发展政策的失败和自然资源的管理不当这两者间密切相关。[1]

可持续发展的概念在20世纪70年代美国环保人士韦斯·杰克逊*的农业著作中首次引起公众的关注。杰克逊认为,应该以最有效的方式对不可再生资源加以利用。[2]

在《我们共同的未来》之前,人们已经讨论过在可用资源有限的情况下人类如何生存这一问题。例如,1972年,罗马俱乐部*(一个由学者、商人和政治家构成的国际智库*)使用计算机建模,制作了一份有影响力的报告《增长的极限》。[3]该报告提供的统计数据,确切表明了未来人口过剩和自然资源需求增长的趋势。作者警告说，除非采取更好、更可持续的方式管理自然资源，否则情况将不可逆转。

> "寻求可持续发展道路的挑战势必会带来动力——确切地说，是强制力——推动我们重新探索多边解决方案并重建国际经济合作体系。这些挑战超越了国家主权的界限，也超越了有限的经济利益策略和相互独立的科学学科的界限。"
>
> —— 世界环境与发展委员会：《我们共同的未来》

学科概览

自18世纪英国人口统计学家托马斯·马尔萨斯*发表关于人口学的著作以来（"人口学"是对社区构成或人口组成的统计研究），人们就开始对有限的环境资源与人类生存之间的关系展开了广泛的讨论。他在《人口论》（1798）一书中指出，人口增长、粮食供应和地球上有限的"为人类提供生存的资源"之间存在联系。[4] 马尔萨斯提出这样一种假设，即世界人口会超越现有的粮食供应能力，因此必须限制人口增长——这与罗马俱乐部报告中提出的问题类似。

在《我们共同的未来》出版之前，人们已经很清楚意识到，必须对地球资源进行可持续管理。这个观点挑战了当时在主流经济学中占主导地位的理论，该理论认为，有限的环境资源和环境退化带来的问题可以通过将人们从世界的一个地区迁移到另一个地区来解决。[5]

学术渊源

早在1972年在瑞典举行的联合国人类环境会议上（也称为斯德哥尔摩会议），就有人试图提出生态可持续性的概念。这是第一次在宏观（即总体和全球）层面评价人类发展对环境的影响并制定保护和改善环境的普遍准则的全球会议。[6] 会议通过了《人类发展宣言》，提出每个人都有权享有健康的环境。各国政府负责执行和维护保护自然环境的政策，联合国正式承认解决世界环境问题的重要性。[7]

同年，罗马俱乐部发表了报告《增长的极限》，明确指出了未来人口过剩和对自然资源需求日益增长的趋势。该报告很受欢迎，许多政治家立即被报告中关于人口过剩与自然资源破坏之间相互联系的观点所说服。然而，后来罗马俱乐部因使用错误数据而受到批评。[8] 布伦特兰委员会委员之一、日本经济学家大来佐武郎*也是罗马俱乐部的成员。

布伦特兰委员会的委员们意识到，围绕全球环境问题的学术和政治辩论仍在持续。作为当时该领域的专家，他们呼吁加强国际合作。例如，委员们认为，全球北方*（主要是北半球的富裕国家）应该为全球南方*（主要是南半球的贫困国家）提供更长期的支持。

---

1. 世界环境与发展委员会：《我们共同的未来》，牛津：牛津大学出版社，1989 年，第 30 页。
2. 韦斯·杰克逊：《农业的新根源》，林肯市：内布拉斯加大学出版社，1985 年，第 144 页。
3. 参见多纳拉·H. 梅多斯等：《增长的极限：罗马俱乐部关于人类困境的研究报告》，纽约：宇宙图书出版社，1974 年。
4. 托马斯·罗伯特·马尔萨斯：《人口论》，伦敦：J. 约翰逊出版社，1798 年，第 13 页。

5. 乔根·诺格尔等:"增长极限的历史",《解决方案》第 2 卷,2010 年第 1 期,第 59—63 页。
6. 费利克斯·多兹等:《只有一个地球:从里约到可持续发展的漫漫长路》,伦敦:劳特利奇出版社,2012 年,第 8—11 页。
7. 多兹等:《只有一个地球》,第 11—12 页。
8. 格雷厄姆·特纳:《增长极限与三十年现实的比较》,联邦科学与工业研究组织工作报告,堪培拉:联邦科学与工业研究组织,2008 年,第 36 页。

3 主导命题

要点

- 《我们共同的未来》的核心问题是如何确保社会发展所必需的经济增长，同时以负责的方式使用和管理有限的自然资源。
- 报告中提到的问题此前曾被其他当代学者讨论过，包括罗马俱乐部、智库、美国环保人士莱斯特·布朗*以及从事环境经济学这一新兴领域研究的学者。
- 报告认为经济增长应优先于自然资源保护，其作者因为这一观点而受到批评；只有一人对批评做出了直接回应。

核心问题

在《我们共同的未来》中，格罗·布伦特兰提出了四个战略目标：

- "提出到 2000 年实现可持续发展的长期环境战略。"
- "加强发达国家和发展中国家之间的合作。"
- "优化解决环境问题的策略。"
- "协助在长期环境问题上达成共识。"[1]

报告的核心问题是如何通过保护环境来确保人类未来的生存和福祉。[2] 在布伦特兰委员会委员们看来，如果不对与自然资源管理相关的国际政策做出重大调整，这个星球现有的生态系统就无法维持人类日益增长的需求。经济危机、全球贫困加剧以及自然的或人为的致命事故等问题表明了有限的环境资源与人类日益增长的需求之间的不平衡。例如，20 世纪 80 年代，印度博帕尔的一家农药工厂和位于今天乌克兰的切尔诺贝利核电站发生了重大工业事故，造成了严重伤亡和对环境的长期负面影响。这些事件迫使国际社会重新评估技术灾难（由技术引起的事故）造成的危害，并且探求更好的国际合作方式，以防止此类事件再次发生。[3]

虽然这份报告的早期版本没有出版发行，但据布伦特兰委员会委员们透露，该报告最初的标题是《受到威胁的未来》。[4] 委员们致力于保护人类免受未来环境和社会危机的威胁。

> "环境不能与人类活动、愿望和需求相割裂而独立存在。企图将环境排除在人类关切的问题之外的做法，使"环境"一词在某些政治场合颇显幼稚。"
>
> —— 世界环境与发展委员会：《我们共同的未来》

参与者

一个被称为"委员会"的国际团队，在布伦特兰的领导下撰写了这份报告。该委员会由 21 位著名学者、政治家和国际官员组成，他们来自阿尔及利亚、巴西、加拿大、中国、哥伦比亚、科特迪瓦、德国、圭亚那、匈牙利、印度、印度尼西亚、意大利、日本、尼日利亚、挪威、沙特阿拉伯、苏联、苏丹、美国、南斯拉夫*和津巴布韦。成员包括加拿大商人、联合国环境规划署（UNEP）*首任主席莫里斯·斯特朗*，印度经济学家尼丁·德赛*，日本前内阁大臣兼罗马俱乐部成员、带领日本实现战后经济增长的大来佐武郎。撰写该报告的 21 名成员被称为"委员"。

该委员会是一个独立的国际机构，旨在提供基于大量定性和定量数据集而进行的全面分析，这些数据集囊括了数字信息和其他类型的研究（如描述和访谈）。为了确保使用最新的信息，委员会与原住民、农民和科学家举行了五次公众听证会。作者们将《我们共同的未来》描述为由许多"各行各业"的人合作撰写的作品。[5]

之前曾有人尝试全面了解如何管理地球日益减少的资源，此类作品包括罗马俱乐部的《增长的极限》，以及美国环保人士莱斯特·布朗的文章和书籍。[6]两者都考虑了管理人口增长和有限自然资源的多种可能情况，有时甚至预测世界末日会到来，认为人们对自然资源的掠夺会最终导致全球灾难。与此相反，布伦特兰委员会相信人类有能力通过可持续发展和平解决这些问题。除了印度经济学家尼丁·德赛最终拒绝接受"可持续发展"这一概念之外，其他每一位参与撰写《我们共同的未来》的团队成员在其后来的工作中都继续使用这一概念。

当代论战

尽管报告提出了我们今天所了解的可持续发展这一概念，但并未解释如何通过管理自然资源来实现可持续性。它采取了以人为本的方式（以人的需求为中心），而不是以自然为中心或以生态为中心的方式（强调自然高于人类需求）。[7]总的来说，《我们共同的未来》支持一种被称为"弱可持续性"*的管理自然资本*（有利于人类繁荣的自然资源）的方式，认为自然资本可能与人力资本*（知识、技术和创新）互换或被人力资本替代。

鉴于此，该报告受到了生态经济学家的严厉批评，他们认为该报告错误地传达出"一种可以放弃环境的安逸感"，[8]似乎实现可持续性发展不需要太多激进的变革。布伦特兰委员会委员们以不同的方式回应了这一质疑。该团队的一名成员尼丁·德赛在批评声中改变了他对可持续发展的看法,后来他对自己曾参与撰写的报告中的部分内容表示反对。[9]

相反,布伦特兰委员会的另一位委员,加拿大商人和环保人士莫里斯·斯特朗仍然坚持报告中的观点。斯特朗认为,制定可持续发展政策的前提是:"人口和人均消费不能超出全球生态系统所能提供的资源限度,这样地球才可能长期满足我们对资源的需求,同时吸收掉产生的废料垃圾。"[10]("生态系统"是指包括存在于特定物理环境中的所有有机体的生物系统。)

---

1. 世界环境与发展委员会:《我们共同的未来》,牛津:牛津大学出版社,1989年,第 viii 页。
2. 世界环境与发展委员会:《我们共同的未来》,第 xi 页。
3. 世界环境与发展委员会:《我们共同的未来》,第 95—235 页。
4. 琳达·斯塔克:《希望之兆:致力于我们共同的未来》,牛津:牛津大学出版社,1990年,第 1 页。
5. 世界环境与发展委员会:《我们共同的未来》,第 xiii 页。
6. 莱斯特·R.布朗:《人、土地和食物:展望未来世界粮食需求》,华盛顿特区:美国农业部,1963年。
7. 帕里尼维尔·毛利:"可持续性与可持续发展",《毛里求斯人报》,2011年8月17日,登录日期2016年2月1日,http://www.lemauricien.com/article/maurice-ile-durable-sustainability-and-sustainable-development。
8. P. A. 维克多等:"弱可持续性有多强?",《应用经济学》第 48 卷,1995年第 2 期,第 75—94 页。
9. 尼丁·德赛:"约翰内斯堡之路研讨会"主旨发言,乔治敦:《环境法评论》,2003年。
10. 莫里斯·斯特朗:《我们究竟要去哪里?》,多伦多:温特吉加拿大出版社,2001年,第 195 页。

4 作者贡献

要点

- 格罗·布伦特兰决心阐明环境与发展是如何相互关联的，并强调我们必须本着负责任的态度管理自然资源。
- 在《我们共同的未来》出版时，可持续发展的理念几乎是一种革命性的经济管理方式，因为它将自然资源考虑在内。
- 布伦特兰先前曾参加一些国际委员会，这使她非常了解有关国际发展的进步思想。

作者目标

1983年12月，联合国秘书长哈维尔·佩雷斯·德奎利亚尔任命格罗·布伦特兰为世界环境与发展委员会主席，自此，《我们共同的未来》的撰写工作开始启动。时任挪威首相的她，将这一任命视为"面向未来、保卫后代人利益的挑战"[1]。布伦特兰之前在发表有关全球经济发展的《勃兰特报告》*的勃兰特委员会和研究有关裁军与全球安全问题的帕尔梅委员会*这两个联合国委员会都有过工作经历，而接受此次任命有助于实现她在之前工作中树立的理想。作为帕尔梅委员会的主席，她旨在实现两个相互关联的主要目标：其一是说服政治领导人重返多边*裁军（即一些国家共同努力并同时实施的政策），尽管由于美国和苏联及其盟友之间正在进行冷战，依然存在核战争的威胁；其二是扩大对人类发展的认识。

虽然布伦特兰对20世纪七八十年代全球合作恶化感到失望，但她仍然相信，如果能在国际层面上提升意识，那么进步思想就有望实现。她对环境退化形势表示担忧，指出这些是人类面临的共同挑战，需要统一的国际应对措施和更公平的全球金融资源分配。在《我们共同的未来》中，环境问题被认为与社会问题密切相关。例如，气候变化和土壤枯竭（农民过度使用造成的土壤贫瘠）是威胁人类生存的复杂问题，发展中（贫穷）国家无法独立解决这些问题，而需要富裕国家的援助。

> "然而，'环境'是我们赖以生存的场所；'发展'是为了在这个场所中改善我们的命运。两者不可分割。而且，就算那些政治领袖们认为自己的国家已经到了其他国家必须努力奋斗才能迎头赶上的高度，他们也必须重视发展问题。"
> —— 世界环境与发展委员会：《我们共同的未来》

研究方法

往届联合国大会，如 1972 年的斯德哥尔摩会议 *，曾讨论过环境问题。这些会议为布伦特兰委员会提供了先例。布伦特兰委员会最重要的成就是认识到自然环境状况与人类福祉之间的联系。布伦特兰希望在《我们共同的未来》中进一步发展这些想法。她对委员会最初的职权范围提出质疑，坚持认为委员会所探讨的不应局限于环境问题；相反，她认为环境问题应该与人类发展相联系。《我们共同的未来》指出，环境与发展实际上是不可分割的。这个报告的主要概念——可持续发展，就是基于这个想法制定的。

根据这份报告，可持续发展受到当前技术发展水平、现有社会机构和地球资源再生能力的限制。报告在研究发展问题时，还考虑了经济增长的因素。可持续发展由环境保护、社会发展（优化人类生活选择）和经济增长三部分构成。

时代贡献

《我们共同的未来》的创新之处在于扩展了对环境问题的界定。布伦特兰和她的同事们认为环境不能"与人类活动、愿望和需求相割裂而独立存在"。他们还认为:"将环境排除在人类关切的问题之外的做法,使得'环境'一词在某些政治场合颇显幼稚。"[2] 联合国第一次正式提出环境与人类发展和经济增长息息相关。

报告的另一个创新之处是提到了后代的需求。大多数联合国的报告是内部文件,只能在联合国机构内部传阅,而《我们共同的未来》与此不同,是作为一本书公开出版。这要归因于报告本身的创新性和向多边国际发展转变的紧迫性。该报告呼吁全世界公民共同努力,实现确保人类生存所需的可持续发展。

---

1. 世界环境与发展委员会:《我们共同的未来》,牛津:牛津大学出版社,1989年,第 ix 页。
2. 世界环境与发展委员会:《我们共同的未来》,第 x 页。

第二部分：学术思想

5 思想主脉

要点

- 《我们共同的未来》明确了 20 世纪人类面临的主要挑战,包括环境危机、贫困、自然资源枯竭、气候变化以及冷战等政治紧张局势。
- 报告将"可持续性"定义为人类"确保其满足当前需要,而不危及后代满足其自身需要"的能力。
- 报告旨在提高政界人士和公众对改变过度消耗自然资源模式的必要性的认识。

核心主题

格罗·布伦特兰的《我们共同的未来》(1987)提出的核心问题是如何在保护环境的同时确保人类在遥远未来的生存和福祉。[1] 1962 年的古巴导弹危机*——可能是美国与其冷战对手苏联最接近核战争的一次对抗,以及一系列严重影响环境的工业事故,如印度博帕尔灾难和苏联切尔诺贝利的核电厂灾难,都对自然环境和人类健康产生了长期的负面影响。政治家和其他西方领导人,包括布伦特兰和她的委员们,都希望在未来预防这种危机并加强国际合作。

该报告的主题是可持续发展迫在眉睫。这一主题的提出是因为人们认识到地球现有的生态系统不足以满足人类日益增长的需求,必须对国际自然资源管理政策进行重大变革。

> "人类活动不能适应这种模式，这从根本上改变着地球系统。许多这样的变化是伴随着威胁生命的危险出现的。我们必须承认这个不可回避的新现实，并加以管理。"
>
> —— 世界环境与发展委员会：《我们共同的未来》

思想探究

《我们共同的未来》意在说服读者相信，可持续发展至关重要。该报告要求政界人士和公众高度重视，并立即采取行动，审查自然资源管理中的不可持续的做法。[2] 该报告提醒人们关注当代人类社会中的重大问题，例如提高公众对环境重要性的认识。这些人们普遍关心的具体问题是我们面临的挑战，是由不可持续的发展趋势造成的。在报告的后面，作者们介绍并讨论了可能的解决方案。

委员会主席格罗·布伦特兰在引言中简要介绍了报告的委托情况和撰写该报告的委员们，随后报告的主体部分立即提出了关键论点："我们对于未来的希望，取决于我们是否现在就采取果断的政治行动来管理环境资源，以确保可持续的人类进步和人类生存。"[3]

该报告开篇的概述部分介绍了人类发展所面临的全球关切和挑战，然后提出国际合作和体制改革形式的可能解决方案。报告主要部分提出的全球关切包括人类的未来生存问题，由贫穷、危机和经济衰退引发的经济崩溃，以及自然资源的过度开发。[4] 报告列举并讨论了全球性挑战，也称为"共同的挑战"，例如人口的快速增长、粮食短缺、气候变化、动植物物种灭绝、能源危机、工业化（工业生产的大量增加）危害和城市发展问题。[5]

报告的结论部分标题为"共同的努力"，讨论了可能的解决方案，提出了需要进行国际合作的领域，例如管理海洋、太空和南极洲等共同的全球资源，以及通过裁军、清洁能源、制度和法律举措来确保全球和平与合作。[6]

语言表述

尽管《我们共同的未来》是由联合国大会委托编写的报告，但其结构与联合国报告的传统格式截然不同，后者通常只提供简短的背景信息，并侧重于拟议的行动。像《我们共同的未来》这样以书的形式出版联合国报告，实乃不寻常之举。

作者们期待该报告可以拥有尽可能多的读者，希望"将其译为年轻人和老年人都可以读懂和领会的语言"。[7] 由于该报告的目标读者是所有人，因此表达观点的语言简截了当。该报告比其他同类题材的报告更全面，不仅阐述了主题，而且试图说服读者相信，可持续发展具有重大意义。

《我们共同的未来》呼吁全球读者高度关注，在人类活动的各个层面立即采取行动。[8] 它表达了当代人类社会的主要关切，如气候变化、粮食短缺、污染和人口快速增长。人们关注的是一系列具体的问题，也是我们面临的"挑战"，是由不可持续的总体发展趋势造成的。报告也介绍和讨论了可能的解决方案。

---

1. 世界环境与发展委员会:《我们共同的未来》，牛津：牛津大学出版社，1989年，第 xi 页。
2. 世界环境与发展委员会:《我们共同的未来》，第 viii 页。
3. 世界环境与发展委员会:《我们共同的未来》，第 1 页。

4. 世界环境与发展委员会:《我们共同的未来》,第 27—36 页。
5. 世界环境与发展委员会:《我们共同的未来》,第 95—257 页。
6. 世界环境与发展委员会:《我们共同的未来》,第 261—343 页。
7. 世界环境与发展委员会:《我们共同的未来》,第 xiv 页。
8. 世界环境与发展委员会:《我们共同的未来》,第 x 页。

6 思想支脉

要点

- 在讨论当前的挑战和可能的解决方案时,《我们共同的未来》使用"因果分析"*来区分危机的原因和后果。
- 这种因果分析法后来发展成为一种新的政策*分析方法,即监测和评估*——在这一过程中,不断检查某项活动以查看其有效性,并随时更新有关信息。
- 基于监测和评估,《我们共同的未来》赞扬非政府组织*是政策变革的主要力量。(非政府组织是独立于其所在国政府的非营利机构。)

其他思想

格罗·布伦特兰在《我们共同的未来》中,通过区分重大危机的原因和后果,将因果分析应用于决策,此举具有开创性意义。在这之前,主要是哲学家,而不是政策制定者,根据原因和影响来解释现象。《我们共同的未来》呼吁进行更深入的分析,带动了其他政策制定者和非政府组织在管理环境资源方面采用这一方法。该报告激发了大量基于监测和评估的研究,据此,研究人员不断评估和跟踪结果。此后,许多经济学家、公共管理者和生物学家(从事生物研究的人员)都制定了方法来评估环境和社会经济政策对生态系统*的重要性和影响。[1]

布伦特兰及其委员们相信人类有能力改变对环境有负面影响的行为模式,并在各个层面开展合作。[2] 报告明确表达了这一信念,并且从该报告的写作背景,以及经历了 1947 至 1991 年冷战时期的几代人的共同愿望中也能在一定程度上看出这一点。这是一个相对乐观的时期，许多人坚信，政府可以共同努力重塑未来。

尽管该报告促进了当时态度和政治思想的改变，但变化的迹象往往被过度乐观地解读。因此，复杂的问题被简化为一个共同的解决方案：国家之间的合作。

> "我们必须更好地了解我们所面临的压力，我们必须找出原因，我们必须设想出管理环境资源和维持人类发展的新方法。"
>
> —— 世界环境与发展委员会：《我们共同的未来》

思想探究

尽管《我们共同的未来》确定了环境与人类发展之间的联系，但是要想了解现有的模式、找到衡量工具、并制定必要的政策干预措施（旨在产生一定效果的政策变化），还有待进一步开展研究工作。不平衡的方法不仅不能解决现有的危机，还有可能导致新的危机。

该报告可能会带来全面的系统性变革，但并没有提供解决实际危机所需要的方法论工具——具体的技术方法。自然环境和人类发展的双重危机是复杂的，难以区分其原因和症状。[3] 该报告以一种非常简单的方式探讨了这些复杂的问题，只是简单地讨论了所面临的挑战，然后提出可能的解决办法。

该报告的因果分析有助于以一种新的方式将原因及其经常出现的非预期影响联系起来。例如，越来越多的工业原料和化学品的使用导致了污染；随着对污染原因的更全面了解，决策者们在减少污染对人类健康和自然环境的负面影响方面做了更好的准备。[4]因果分析还具有教育意义，因为它有助于说服读者以个人身份去采取更加可持续的做法。

被忽视之处

虽然《我们共同的未来》及其主要思想已经得到充分研究，但报告中有关非政府组织在促进可持续发展中的作用的见解，在当前的研究中仍然被忽视，只有少数文献涉及。[5]非政府组织与可持续发展之间的关系并不明显。事实上，如印度活动家范达娜·席瓦*等激进的环保主义者甚至抱怨说，可持续发展对环境运动有害，特别是在相对贫穷国家，因为它有可能扼杀更激进的环保斗争。[6]她意识到，西方国际组织引入"弱可持续性"可能会掩盖非政府环保组织游说的"强可持续性"*。〔"弱可持续性"是可持续性的一种形式，认为自然资本（资源）可能被人力资本（知识、技术等）替代；"强可持续性"就是优先考虑环境的可持续性。〕

许多非政府组织为《我们共同的未来》的研究做出了贡献，提交了他们收集的事实并提出了有关政策变革的建议。作者们知道，政府接受政策变革并不容易，因为可持续发展的实践需要对资源管理、技术、制度和政策进行根本性的变革。这样的变革很难实施，需要制度和财政资源，这就要求政治家们做出"痛苦的选择"[7]。

从一开始，《我们共同的未来》的作者们就预料到，可能会遇到对激进变革的抵制。他们预计，政府在解决环境危机方面会面临严重困难，因为他们的关注点往往"过于狭窄，过于注重生产或增长数量"[8]。从这个意义上说，可持续发展是一个"非政府"议题——也就是说，政府不一定是推动社会可持续发展的最佳选择，需要外部压力和鼓励。在布伦特兰的愿景中，非政府组织是提高公众意识和推动政府行动的先锋。[9] 然而，该报告的委员们从来没有讨论过非政府组织的弱点：它们往往关注特定的利益，而且在没有政府支持的情况下，为项目提供资金的能力有限。

---

1. 威廉·R. 沙迪什等：《项目评估的基础：实践理论》，伦敦：塞奇出版社，1991年，第20—21页。
2. 世界环境与发展委员会：《我们共同的未来》，牛津：牛津大学出版社，1989年，第 ix 页。
3. 大卫·沃斯戴尔：《全球动力学研究七——布伦特兰及其他：走向全球化进程》，伦敦：厄奇出版社，1987年，第7—8页。
4. 世界环境与发展委员会：《我们共同的未来》，第2—3页。
5. 本·派尔："祝'绿色和平'生日不快乐"，《尖刺》，2011年9月12日，登录日期2016年2月2日，http://www.spiked-online.com/newsite/article/11068#.VrCGxPkS-Uk。
6. 范达娜·席瓦："全球范围的绿化"，载《全球生态：政治冲突的新舞台》，沃尔夫冈·萨克斯编，伦敦：泽德出版社，1993年，第149—156页。
7. 世界环境与发展委员会：《我们共同的未来》，第9页。
8. 世界环境与发展委员会：《我们共同的未来》，第328页。
9. 世界环境与发展委员会：《我们共同的未来》，第328页。

7 历史成就

要点

- 《我们共同的未来》的重要性在全世界都得到了认可；报告提出的建议仍然主导着联合国的政治议程。
- 报告阐述了普世的价值观，赢得了广大读者的认可。
- 但是，报告未能直接引发重要政策变革或个人行为的重大变化。

观点评价

顾名思义，格罗·布伦特兰的《我们共同的未来》，旨在提出跨越国家和时代的普遍诉求。它的目的是改变国际和国家政策，以及破坏环境的个人行为，将之作为"全球变革议程"[1]的一部分。除了报告中讨论的问题是全球性的以外，21位来自不同国家的委员还"一致认为安全、福祉和地球的生存取决于我们着手进行的变革"，[2]因此，该报告与全世界息息相关。

《我们共同的未来》着眼于解决国际社会在20世纪80年代面临的主要挑战。1987年该报告发表时，作者们希望能在2000年前解决这些挑战。[3]但是，决策者和公众对报告的反响不足，以致于未能及时实现目标。报告中提到的所有问题，如污染、人口增长、经济危机、贫困和气候变化等，如今仍然存在，甚至随着时间的推移而恶化。因此，报告中讨论的问题对今天的读者来说仍然是非常熟悉的。

《我们共同的未来》指出了问题的全球性，以及它们如何适用于发达国家和发展中国家，即富裕国家和贫穷国家。[4]但报告也指出，发展中国家在知识和经济资源等方面的能力极其受限，并号召发达国家增加对贫穷国家的援助。⁵ 一些学者和活动家,如瑞士经济学家马塞厄斯·芬格*,批评了报告中对富有的发达国家和贫穷的发展中国家的区分。他们认为这样的区分削弱和分裂了环境("绿色")运动。⁶

> "我记得,对目前的结果我给出了一种较为冷静的警告,我承认,我们在许多领域都有进步,我们在一些领域有一些进步,但也在一些领域毫无进展。"
>
> ——格罗·布伦特兰,接受波罗的海大学彼得·奥克斯凯采访,
> 瑞典乌普萨拉,1997

当时的成就

在《我们共同的未来》出版三年后,加拿大环境部长吕西安·布沙尔*表示,报告中提出的可持续发展的概念"永远改变了我们对环境的看法"⁷。

报告发表数十年以来,可持续发展的概念已被广泛接受,并成为决策者的重要参照。可持续发展理念不仅催生了环境经济学这一新学科,⁸ 也在各个层面上影响了决策过程,并成为通用的政治术语。

到2009年,可持续发展议程纳入了106个国家的国家发展战略。企业界接受了一项新议程,即企业社会责任*,尽管一些企业机构因有"漂绿"*之嫌而遭到批评。所谓"漂绿",是指企业和机构将自身包装成环保友好型的营销手段。⁹ 对造成环境破坏的商业行为的批评,促使企业和环境与社会方面的非政府组织合作开展联合行动来解决可持续发展问题。各国政府在达成约束性(强制性)

协议方面仍进展缓慢,这些协议涉及为污染、气候变化和粮食短缺等全球事务共同承担责任。

报告促进多边合作的目标已经部分实现。从这个角度来看,这份报告改变了国家和国际政策,这远远超出了作者们的预料。

局限性

20世纪80年代,布伦特兰领导的世界环境与发展委员会希望在2000年之前实现可持续发展。[10] 然而,尽管《我们共同的未来》提出了行动建议,但报告中提到的问题迄今尚未解决,因为"发展与环境之间的紧张关系、争议和僵局依然存在"。[11]

尽管可持续发展的概念已得到了全世界的认可,但在实现这一概念方面进展缓慢。《我们共同的未来》所提出的措施是自愿实施的。[12] 该报告的制度跟进,包括1992年在里约热内卢召开的联合国地球峰会*和2012年"里约+20"峰会*,在"构思从理论到实践的转变"方面面临严重困难。[13] 主要问题在于,"尽管世界各地有大批支持者在深度关注并热议可持续发展,但都没有采取严肃的实际行动"。[14] 政策分析人员对1997年《京都议定书》*(旨在通过减少某些被认为会导致气候变化的气体排放,来减缓气候变化的国际条约)和2015年举行的关于气候变化问题的巴黎会议*有着类似的感受。[15]

在理论层面上,可以将可持续发展概念分为三个部分(经济、社会和环境发展),但这又带来了另一项挑战:如何在环境和经济之间找到平衡点。这个挑战引发了关于"两者应该优先考虑谁"这一基本问题的讨论:是经济增长更重要,如弱可持续发展模式,还是环境保护更重要,如强可持续发展模式?[16]

1. 世界环境与发展委员会:《我们共同的未来》,牛津:牛津大学出版社,1989 年,第 ix 页。
2. 世界环境与发展委员会:《我们共同的未来》,第 343 页。
3. 世界环境与发展委员会:《我们共同的未来》,第 ix 页。
4. 世界环境与发展委员会:《我们共同的未来》,第 52 页。
5. 世界环境与发展委员会:《我们共同的未来》,第 60 页。
6. 马塞厄斯·芬格:"联合国环境与发展会议进程中的政治",载《全球生态:政治冲突的新舞台》,沃尔夫冈·萨克斯编,伦敦:泽德出版社,1993 年,第 36 页。
7. 琳达·斯塔克:《希望之兆:致力于我们共同的未来》,牛津:牛津大学出版社,1990 年,第 21 页。
8. 赫尔曼·戴利:《生态经济学与可持续发展:论文选集》,切尔滕纳姆:爱德华·艾尔加出版社,2007 年,第 251 页。
9. 约翰·德雷克斯哈格和黛博拉·墨菲:《可持续发展:从布伦特兰到里约 2012》,纽约:联合国,2010 年,第 15 页。
10. 世界环境与发展委员会:《我们共同的未来》,第 ix 页。
11. 沃尔克·豪夫:"布伦特兰报告:20 年更新"主题演讲,欧洲可持续发展,柏林,2007 年 6 月 3 日,登录日期 2016 年 2 月 1 日,http://www.nachhaltigkeitsrat.de/uploads/media/ESB07_Keynote_speech_Hauff_07-06-04_01.pdf。
12. 珍妮佛·艾略特:《可持续发展导论:发展中的世界》,伦敦:劳特利奇出版社,2000 年,第 8 页。
13. 艾略特:《可持续发展导论》,第 8 页。
14. 德雷克斯哈格和墨菲:《可持续发展》,第 2 页。
15. 托马斯·斯特纳:"巴黎气候变化会议需要更雄心勃勃",《经济学人》,2015 年 11 月 18 日,登录日期 2016 年 2 月 2 日,http://www.economist.com/blogs/freeexchange/2015/11/cutting-carbon-emissions。
16. 德雷克斯哈格和墨菲:《可持续发展》,第 10 页。

8 著作地位

要点

- 格罗·布伦特兰为当代全球挑战的讨论做出了重要贡献,这些挑战包括环境退化、气候变化和人类健康。
- 《我们共同的未来》在20世纪80年代为人类发展提供了一种革命性的方法,尽管最近受到批评,但它仍然是可持续发展方面最有影响力的成果;格罗·布伦特兰一直致力于推动可持续发展。
- 《我们共同的未来》是所有布伦特兰委员会委员的职业生涯中一项里程碑式的成果,包括布伦特兰本人。

定位

格罗·布伦特兰应邀担任世界环境与发展委员会主席时,已经是一位经验丰富的政治家。她不仅是挪威第一位女首相,连任两届(1986—1989和1990—1996),还是唯一担任过环境大臣的首相(1974—1979)。

这一独特的背景使布伦特兰能够在管理有限的资源和环境以及人类发展方面提出新的想法。在《我们共同的未来》出版之前,布伦特兰已经享誉盛名,而委员会的报告提高了她作为名人政治家的地位。她随后又应邀撰写短文,记录委员会历史上的重要时刻,包括为一本介绍报告编写情况的书撰写前言,并做了许多公开演讲。然而,从那之后,她就再没有创作出可以与《我们共同的未来》相提并论的重要作品。[1]

在1997至1998年间,布伦特兰用挪威语写了两部自传作品,书名分别为《我的人生》[2]和《难忘的十年 1986—1996》[3]。2002年，她用英文写了一部题为《女首相：权力与政治生活》[4]的自传。这本书记录了她生命中的重要时刻，以及她与其他杰出领导人的接触，如法国总统弗朗索瓦·密特朗*、法国政治家雅克·德洛尔*、俄罗斯联邦首任总统鲍里斯·叶利钦*、美国政治家希拉里·克林顿*以及南非自由战士和总统纳尔逊·曼德拉*。她的丈夫阿恩·奥拉夫·布伦特兰也是挪威的政治家，他写了两本回忆录，记录了格罗·布伦特兰执政期间他们的生活。书名分别为《娶妻格罗》和《仍然娶妻格罗》。[5] 布伦特兰担任联合国世界卫生组织总干事期间，撰写了关于环境因素和气候变化对人类健康影响的报告。最近她公开讨论了民主和人权在实现可持续发展中的作用。

> "最终，我知道接受联合国给我的挑战是正确的选择。它对这个国家（挪威）和全球产生了巨大的影响。"
>
> —— 格罗·布伦特兰，接受乌贡姿研究所采访

整合

布伦特兰一直支持《我们共同的未来》的主要观点：环境资源管理与经济增长之间存在着重要的联系。她认为北半球的发展问题虽然已基本解决，但她对发展中国家的大规模贫困表示担忧，自报告发表以来，这种情况似乎还是在恶化。报告的知名批评家也经常提到这一点，如印度环境活动家范达娜·席瓦[6]和瑞士经济学家马塞厄斯·芬格。[7]《我们共同的未来》讨论了人类对自然的普遍责任以及全球北方（北半球发达国家）与全球南方（南半球发展中国家）合作的必要性；布伦特兰后来在文章中进一步指出，北半球的

消费者应对全球南方的贫困和环境退化承担部分责任，而这可能会导致未来的冲突。[8]

布伦特兰后来在世卫组织工作期间，在其文章中采用了与《我们共同的未来》类似的方法。1998年，她与经济学家杰弗里·D.萨克斯*合作发表了"宏观经济与健康"一文，该文指出，健康与长寿既是发展的根本目标，也是实现发展目标的手段，并引用了可持续发展的概念。[9]世界卫生组织的报告还将健康与减贫以及长期经济增长联系起来。

可以说，布伦特兰一直坚持她在报告中提出的观点，认为经济增长是实现国际发展目标的手段。报告没有讨论人类经济活动对人类发展造成的负面环境影响——此后布伦特兰也再未提及。

意义

《我们共同的未来》是布伦特兰迄今为止最著名的作品。她开始主持委员会工作并负责组织撰写该报告时，已经在联合国界享有盛誉，这一点提高了其委员会的政治可信度。因为负责制定该报告，并且提出了可以说是现代政治经济最重要的概念之一——可持续发展，布伦特兰因此享誉全球。

如今，身为长老会*（一个由退休的联合国高级官员和前政治领导人组成的智库）的成员，布伦特兰一直坚持《我们共同的未来》中提出的观点和理想。她在后来出版的不太为人所知的著作中重申了这些观点，并在报告发表后的几年里，特别是在她担任世界卫生组织总干事期间，一直在推行可持续发展的概念。

虽然《我们的共同未来》中提出的新观点在20世纪80年代得到了大多数政府和公众的欢迎[10]，但近期对该报告及其核心概念可持续发展的解读却更多地带有批判性。环保活动家和政治经济学家把这份报告称为新自由主义思想的一个例子，认为它主张通过技术进步和国际合作从而有效利用资源，以实现经济无限增长，这本身就是不可能的。[11]（"新自由主义"*是经济学的一种研究方法，它要求市场在不受监管或没有政府干预的影响下运作，而不考虑任何社会后果。）尽管布伦特兰在2013年的维也纳公开演讲中承认并接受了这一批评，但她并没有就此发表过任何文章。

---

1. 参见琳达·斯塔克：《希望之兆：致力于我们共同的未来》，牛津：牛津大学出版社，1990年。
2. 格罗·哈莱姆·布伦特兰：《我的人生：1939—1986》，奥斯陆：吉伦哈尔出版社，1997年。
3. 格罗·哈莱姆·布伦特兰：《难忘的十年 1986—1996》，奥斯陆：吉德达尔出版社，1998年。
4. 格罗·哈莱姆·布伦特兰：《女首相：权力与政治生活》，纽约：法勒、施特劳斯和吉鲁出版社，2005年，第19页。
5. 参见阿尔纳·奥拉夫·布伦特兰：《娶妻格罗》，奥斯陆：奥斯陆大学出版社，1996年；阿尔纳·奥拉夫·布伦特兰：《仍然娶妻格罗》，奥斯陆：奥斯陆大学出版社，2003年。
6. 范达娜·席瓦："全球范围的绿化"，载《全球生态：政治冲突的新舞台》，沃尔夫冈·萨克斯编，伦敦：泽德出版社，1993年，第149—156页。
7. 马塞厄斯·芬格："联合国环境与发展会议进程中的政治"，载《全球生态：政治冲突的新舞台》，沃尔夫冈·萨克斯编，伦敦：泽德出版社，1993年，第36页。
8. 格罗·哈莱姆·布伦特兰："全球变化与我们共同的未来：本杰明·富兰克林讲座"，载《全球变化与我们共同的未来》，R. S. 德弗里斯和T. F. 马隆编，华盛顿特区：国家科学院出版社，1989年，第15页。
9. 格罗·哈莱姆·布伦特兰和杰弗里·D.萨克斯："宏观经济与健康：经济发展的健康投资"，日内瓦：世界卫生组织，2001年，第3页。
10. 斯塔克：《希望之兆》，第2页。
11. 格里德·奥·图泰尔等编：《地缘政治读本》，伦敦：劳特利奇出版社，1998年。

第三部分：学术影响

9 最初反响

要点

- 尽管《我们共同的未来》及其主要概念——可持续发展——获得普遍肯定,却也受到了批评。批评主要来自环保活动家和经济学家,他们批评该报告采用了一种被称为"弱可持续性"的方法。
- 报告的作者们都没有与其批评者进行公开辩论。
- 格罗·布伦特兰在后来的文章中对报告中给出的可持续发展三要素模式进行了补充,引入了第四要素——文化,这在一定程度上是对批评作出的回应。

批评

虽然格罗·布伦特兰的《我们共同的未来》中有关提高环境意识的建议受到了科学家和决策者的普遍欢迎,但可持续发展这一新概念的某些方面却受到了批评。最早的批评来自环保活动家和经济学家,他们认为报告中描述的可持续发展概念过于宽泛,并没有具体说明必须可持续发展什么。在报告中,可持续性既用于指负责任地管理自然资源,也用作经济术语。例如,作者在提及外债时就用到了可持续性概念,当外债风险导致发展中国家产生资金外逃(较不富裕的国家的资金被投资到其他地方)这一现象时,就被描述为"不可持续的"[1]。通常,在报告中,可持续发展等同于可持续增长,意指在负责任地管理自然资源的新秩序下有望实现经济繁荣。[2]

印度社会科学家希夫·维斯瓦纳坦*质疑将环境可持续性与发展二者结合的现实性。他认为,联合国等国际组织强加的发展模式是对发展中国家施加的"种族灭绝的控制行为",并没有充分解决真正的生态问题。[3]

《我们共同的未来》将环境状况与经济发展联系起来的方式在当时具有创新性。但生态经济学家称这种方法为"弱可持续性",因为它将创新和知识视为自然资本(对人类生产生活有用的自然资源)的可能替代品。[4] 对他们来说,这种方法给人一种"可以放弃环境的感觉"。相反,他们要求"强可持续性"——一种更加强调保护自然资源的方法。[5]

> "在我最后的回答中,我说:'是的,万物相关。'20年前,我作为首相发表的最后一份报告立即遭到反对派的批评甚至嘲笑,认为它含糊其辞、表达不清。有趣的是,今天人们常常以极大的敬意甚至钦佩之情引用这份报告。"
>
> —— 格罗·布伦特兰:"气候变化与我们共同的未来",在全球气候与能源项目研讨会上的讲话

回应

布伦特兰一直坚持她在报告中提出的关于可持续发展必要性的立场。多年来,她把民主和人权问题也涵盖在内,进一步深化了对可持续性概念的理解。[6] 有人批评该报告是一种漂绿行为(对生态问题的表面支持,而不是进行真正的改变,通常以物质利益为目的),会破坏全球环境运动,如印度环保活动家范达娜·席瓦所言。[7] 然而,布伦特兰和其他参与撰写报告的委员都没有就此事与其批评者展开公开辩论。

《我们共同的未来》讨论了人类保护自然的共同责任以及全球北方和全球南方之间合作的必要性。在后来的文章中,布伦特兰又

进一步提出，北方消费者加剧了全球南方的贫困化，这可能成为未来冲突的根源。[8] 这可以算是她对批评者的回应。

有生态经济学家批评可持续发展这一提法，不同意报告中提出的关于自然资本可以被知识、技术和创新等"人力资本"取代的观点。布伦特兰在多大程度上赞同他们的观点还不得而知。似乎没有人想过直接问她这个问题。众所周知，她是一位谨慎的政治家，不会轻易选择一个公共平台或者随意用文字来明确表达自己的观点。[9]

冲突与共识

《我们共同的未来》中提出的可持续性概念受到经济学家和环保活动家的批评，因为报告提出促进经济增长比环境保护更重要。该报告还因为采用了新自由主义发展模式而受到批评，根据该模式，市场必须在不受阻碍的情况下运行，全球北方可以向全球南方发号施令，无视或破坏当地的环境运动。[10]

然而，报告的作者认为，他们的主要目标是提高人们对环境和社会的危机意识，而不是推行一套确切的行动。和布伦特兰一样，许多作者本身就是政治家，他们的地位使他们不能对世界秩序提出激进的改革方案。事实上，他们认为可持续发展理念已经足够激进，并意识到进一步推动改革所涉及的政治风险。

《我们共同的未来》和布伦特兰的演讲表明，可持续性的概念是以人类及其需求为中心的，她不认为这有什么不对。人类和自然被视为一个相互联系的系统："人类影响着决定地球命运的趋势。地球也影响着人类。"[11] 布伦特兰认为，关注人类需求对人类和地球都有好处。

布伦德兰在回应对该报告的批评时做出的主要妥协是，在可持续发展的三大支柱体系中引入第四个要素：文化。[12] 她为可持续性概念补充了文化相对主义的概念，即不同的文化有不同的价值观和实践，不能直接相互比较。据此，她就可以解释为什么不同的国家和地区会以不同的速度和方式引入可持续性。该报告曾被批评为"一刀切"，而这个新增加的文化要素使得该报告避免了这一批评。

---

1. 世界环境与发展委员会：《我们共同的未来》，牛津：牛津大学出版社，1989年，第73页。
2. 世界环境与发展委员会：《我们共同的未来》，第68页。
3. 希夫·维斯瓦纳坦："布伦特兰夫人的幻灭宇宙"，《选择杂志》第16卷，1991年，第378—381页。
4. 大卫·皮埃尔斯和吉尔斯·阿特金森："可持续发展的概念：布伦特兰报告十年功用评估"，《瑞士经济与统计杂志》第134卷，1998年第3期，第4—5页。
5. P. A. 维克多等："弱可持续性有多强？"，《应用经济学》第48卷，1995年第2期，第75—94页。
6. 格罗·哈莱姆·布伦特兰：《2015年诺贝尔和平奖论坛开幕致辞》，奥斯陆，2015年3月10日，登录日期2016年2月1日，https://www.youtube.com/watch?v=LBRmSjsnVGs。
7. 范达娜·席瓦："全球绿色生态"，载《全球生态：政治冲突的新舞台》，沃尔夫冈·萨克斯编，伦敦：泽德出版社，1993年，第149—156页。
8. 格罗·哈莱姆·布伦特兰："全球变化与我们共同的未来：本杰明·富兰克林讲座"，载《全球变化与我们共同的未来》，R. S. 德弗里斯和T. F. 马隆编，华盛顿特区：国家科学院出版社，1989年，第15页。

9. 大卫·威尔斯福德:《当代西欧政治领袖:传记字典》,康涅狄格州西港:格林伍德出版社,1995年,第55页。
10. 维斯瓦纳坦:"布伦特兰夫人的幻灭宇宙",第378—379页。
11. 格罗·哈莱姆·布伦特兰:"健康人类,健康星球",商业与环境年度讲座,伦敦,2001年3月15日,登录日期2016年2月1日,http://www.cisl.cam.ac.uk/publications/archive-publications/brundtland-paper。
12. 格罗·哈莱姆·布伦特兰:"ARA讲座",维也纳技术大学,2013年11月18日,登录日期2016年2月1日,https://www.youtube.com/watch?v=X7Z2o8tZZoE。

10 后续争议

要点

- 环保活动家和经济学家对《我们共同的未来》和可持续性的概念提出了批评，认为该报告过分关注经济增长。
- 联合国官员和布伦特兰委员会委员们已经认识到在实现可持续发展方面进展缓慢。
- 来自发展中国家的新一代积极分子和进步经济学家提供了针对报告内容的替代性构想。

应用与问题

《我们共同的未来》是在格罗·布伦特兰的领导下合作研究和撰写的。其核心概念可持续发展，自发表以来一直是当代关于人类发展辩论的核心话题；它还是众多当代全球事务的主要分析框架。该报告仍然有助于提醒现任领导人注意自然环境与国际发展之间的复杂关系。

2012 年，联合国在巴西里约热内卢举办了可持续发展会议——"里约+20"峰会，各国政府在会议上达成了一项非约束性协议，名为《我们想要的未来》。虽然一些国家政府主动承诺实现协议所规定的可持续发展，但该文件缺乏强制性承诺，使人们普遍对会议的结果感到失望；媒体和民间团体认为这是"讽刺外交的漫画"。[1]

当前国际发展问题严重，而在解决这些问题方面又缺乏进展，这促使人们改变了对《我们共同的未来》的看法。对于该报告及报告中关于世界成功应对经济、环境和社会危机能力的乐观描述，人们的总体态度已从最初的赞扬和乐观转变为更具批判性、有时甚至悲观的观点。有些人士，如斯洛文尼亚社会理论家德拉戈·科斯*，对世界是否有能力实现可持续发展表示了强烈的怀疑。[2]

> "虽然可持续发展往往被视为环境问题，但它也一直受到各种不同议程的影响。"
> —— 约翰·德雷克斯哈格和黛博拉·墨菲：
> 《可持续发展：从布伦特兰到里约 2012》

思想流派

最初，所有参与《我们共同的未来》撰写工作的人都一致认为，我们必须可持续地管理环境资源。后来，支持强可持续性和弱可持续性的两方展开辩论，前者强调环境保护，后者侧重经济发展。[3] 弱可持续性的支持者包括布伦特兰委员会委员，如德国政治家沃尔克·豪夫*和加拿大商人、外交官莫里斯·斯特朗。斯特朗认为，经济增长乏力导致了环境恶化。这种方法往往把经济增长放在首位，自然得到政府和企业的支持。

相反，环境或生态经济学家，如强可持续性的支持者赫尔曼·达利*认为，应优先保护自然资源，否则将导致经济进一步恶化。生态经济学学派接受了强可持续性和弱可持续性的概念，并将其纳入关于自然资本在资源管理中的地位的讨论。[4]

当代研究

环境经济学家和环保活动家认为，强可持续性优于弱可持续性，这一观点仍存在争议。

关于强可持续性的观点，有一个最新提法，名为"去增长"，意思是拒绝新自由主义的经济增长理念。这个概念是由草根环保运动提出的，在南美洲尤为盛行，由哥伦比亚裔美国人类学家阿图罗·埃斯科瓦尔*首次引入学术框架中。这个概念重新阐释了经济增长的含义。在去增长模式下，经济只能在健康和教育等造福人民或惠及民生的领域增长。[5] 这一模式提倡强可持续性方法，因为它主张保护自然。2010年4月在玻利维亚科恰班巴提出的《地球母亲权利世界宣言》，[6] 就是这种模式得以践行的一个例子。这是在1992地球峰会上未能通过的《地球宪章》的替代方案。[7] 玻利维亚率先通过了该文件，该文件将地球权利的概念列入联合国议程。

生态经济学家也持续对以生产和消费为基础的经济增长模式展开批评，特别是当它超过了自然资源自我补充的速度时。例如，哈佛大学经济学家迈克尔·波特*[8] 设想，在技术创新的基础上更有效地利用资源，这种双赢的局面避免了过度开发自然资源。这一观点与《我们共同的未来》一致，认为技术是应对全球挑战的解决方案之一。基于波特的研究，新一代经济学家[9] 为建设可持续经济提供了一个全面的框架。该框架包括以伦理价值为基础的新商业模式，较少关注经济利润，促进以社区为基础的所有权结构，如合作社（由其成员所拥有的企业或服务），并旨在制定更好的社会经济进步措施（社会和经济状况好转时的措施）。

---

1. 吉姆·利普："这正在发生，但不是在里约"，《纽约时报》，2012年6月24日，登录日期2016年2月1日，http://www.nytimes.com/2012/06/25/opinion/action-is-

happening-but-not-in-rio.html?_r=0。
2. 德拉戈·科斯：“可持续发展：实现乌托邦？”，《社会学杂志》第 54 卷，2012 年第 1 期，第 7—20 页。
3. 杰罗姆·佩伦克：“弱可持续性与强可持续性”，GSDR 简报，鲁汶：鲁汶大学出版社，2015 年，第 1—4 页。
4. 约翰·M. 高迪和玛莎·沃尔顿：“生态经济学中的可持续性概念”，《经济学与其他学科的互动》，第 2 卷，《生命支持系统百科全书》，巴黎：教科文组织／生命支持系统百科全书，2008 年，第 111—120 页。
5. 阿图罗·埃斯科瓦尔：“发展选择”，接受罗布·霍普金斯采访，威尼斯，2012 年 9 月 28 日，登录日期 2016 年 2 月 1 日，http://transitionculture.org/2012/09/28/alternatives-to-development-an-interview-with-arturo-escobar/。
6. 全球自然权利联盟：“地球母亲权利世界宣言”，玻利维亚科恰班巴，2010 年 4 月 22 日，登录日期 2015 年 12 月 14 日，http://therightsofnature.org/universal-declaration/。
7. 莫里斯·斯特朗：“1992 年地球峰会：内部视角”，接受菲利普·沙别科夫采访，魁北克省，1999 年，登录日期 2016 年 2 月 1 日，http://www.mauricestrong.net/index.php/earth-summit-strong。
8. 迈克尔·E. 波特和克拉斯·范德林德：“绿色和竞争：结束僵局”，《哈佛商业评论》第 73 卷，1995 年第 5 期。
9. D. W. 奥尼尔等：“够了：在资源有限的世界中坚持可持续经济的想法”，《稳态经济会议报告》，利兹：稳态国家经济促进中心和全民经济正义，2010 年，第 9 页。

11 当代印迹

要点

- 《我们共同的未来》在学术思想和公共政策方面仍然意义重大。
- 《我们共同的未来》中提出的可持续性概念受到了批评,因其在实践中难以实现。
- 格罗·布伦特兰仍然在国际政治领域担任高级政策顾问,并一直致力于推广可持续发展的理念。

地位

格罗·布伦特兰的《我们共同的未来》是可持续发展概念的明确出处,并且是当前关于未来发展路径辩论的核心。[1]

实现可持续发展的主要障碍是,需要采取的行动非常零碎,不仅涉及各国政府行为以及政府间协同合作方式的改变,而且涉及个人的行为的改变。另一个障碍是可持续发展概念的模糊性,及其所涵盖问题的复杂性。[2] 关于人类发展的未来,这个话题利害攸关;因此,关于可持续发展的辩论早已经不仅限于那些对生态经济学感兴趣的人,而是一个关乎所有人的问题。

联合国作为最初委托撰写《布伦特兰报告》的机构,从未质疑过可持续发展的可行性。相反,它试图接受批评并调整这一概念以适应新的挑战。联合国的机构承担了将可持续发展付诸实践的任务。该报告之后首次在制度上的后续跟进是在 1992 年的地球峰会上,联合国官员们,包括布伦特兰委员会的成员、大会秘书长莫里斯·斯特朗,把必要的政策变革落实为一套具体的行动,即《21世纪议程》*。关于推广可持续发展的另一个尝试是在20年后，里约热内卢举办的"里约+20"峰会上，这次会议达成了一项国际协议，即《千年发展目标》*，是全球范围内人类发展的目标清单，其中包括可持续性目标，并被称为"可持续发展目标"（SDGs）。

> "可持续发展是一种富有远见的发展模式，过去20年来，政府、企业和民间组织已将可持续发展作为指导原则……"
>
> ——约翰·德雷克斯哈格和黛博拉·墨菲：
> 《可持续发展：从布伦特兰到里约2012》

互动

联合国已举办数次会议以推动实施《我们共同的未来》所提出的可持续发展。其中，1992年的"地球峰会"和2012年的"里约+20"峰会尤为重要。这两次峰会遭到批评，被认为过分强调环境问题，而对富裕国家向贫困国家提供发展援助以及国家之间的合作关注太少。[3]

格罗·布伦特兰和莫里斯·斯特朗都是布伦特兰委员会的知名成员，他们20多年来一直致力于探讨可持续发展。尽管有人对可持续发展进展缓慢这一点提出了严厉批评，但布伦特兰和斯特朗仍然坚持这一理念，认为可持续发展面临的阻碍并不代表这一概念有根本缺陷，而要归因于其他因素。

在布伦特兰看来，可持续发展面临的挑战要比实施它的系统面临的挑战更大。她在最近的发言中提到，国际机构可以在实现可持续性目标方面取得更好的进展。斯特朗也指出了联合国的体制弱点，这些弱点使得联合国在可持续性方面不具备强有力的全球领导力。但他高度肯定1992年至1996年间联合国秘书长布特罗斯·布特罗斯－加利*为改革联合国所作的努力，"整合秘书处内的若干资源，组建一个新的经济和社会事务部门"，被称为经济及社会理事会（ECOSOC）*，这是一个负责经济和社会活动的机构。[4]该理事会在落实里约峰会所作出的决定方面发挥了核心作用。

斯特朗曾在"里约+20"峰会上担任高级顾问，并借此机会建议对联合国系统进行制度改革；其中包括重组联合国环境规划署，该署在联合国系统内负责协调机构的环境活动和政策，并向力图实施环保行动和政策的发展中国家提供援助。[5]尽管受到批评，但"里约+20"峰会成功地将"千年发展目标"重新定为"可持续发展目标"，确保将可持续发展目标纳入官方的国际发展目标。这是一项重要成果。

持续争议

关于如何使可持续性在实践中发挥作用的争论仍在继续。虽然有些人，包括斯洛文尼亚的社会科学家德拉戈·科斯，[6]认为这个概念是乌托邦式的（也就是说，它属于一个并不存在的完美世界），但大多数科学家和政治家认为可持续发展有可能成为现实。包括南非宗教领袖和活动家德斯蒙德·图图*在内的许多人则坚持认为，如果人类想在不久的将来生存下去，就必须付出额外的努力。

联合国官员和布伦特兰委员会的委员们都意识到在实施可

持续性发展方面进展缓慢。2010年，国际可持续发展研究所＊（一个旨在通过可持续方式来促进人类发展的独立的、非营利组织）的科学家受托组成了联合国全球可持续发展高级别小组来讨论这个问题，研究进展缓慢的原因和可能的解决方案。某些发展专家认为，为了实现可持续发展，全球经济——尤其是在全球商业领域——必须发生深刻的体制变革。[7]

布伦特兰本人也是联合国全球可持续发展高级别小组成员，她也提到了这个问题。在维也纳科技大学的一次公开演讲中，她回答了以下问题："里约会议（1992年地球峰会）至今已过去了20年，我们为什么还不能改变我们的行为方式并建立一个可持续的未来？"[8] 她认为之所以进展缓慢，有政治、社会和技术方面的原因，但主要原因在于全球政府体系的低效，这导致未能广泛实施可持续发展。[9]

---

1. 沃尔克·豪夫："布伦特兰报告：20年更新"主题演讲，欧洲可持续发展，柏林，2007年6月3日，登录日期2016年2月1日，http://www.nachhaltigkeitsrat.de/uploads/media/ESB07_Keynote_speech_Hauff_07-06-04_01.pdf。

2. 约翰·德雷克斯哈格和黛博拉·墨菲：《可持续发展：从布伦特兰到里约2012》，纽约：联合国，2010年，第16页。

3. 莫里斯·斯特朗：《我们究竟要去哪里？》，多伦多：温特吉加拿大出版社，2001年，第78页。

4. 莫里斯·斯特朗："1992年地球峰会：内部视角"，接受菲利普·沙别科夫采访，魁北克省，1999年，登录日期2016年2月1日，http://www.mauricestrong.net/index.php/earth-summit-strong。

5. 莫里斯·斯特朗："莫里斯·F.斯特朗在里约+20的特别联合国大会活动的发

言",纽约,2011 年 10 月 25 日,登录日期 2016 年 2 月 1 日,http://www.unep.org/environmentalgovernance/PerspectivesonRIO20/MauriceFStrong/tabid/55711/Default.aspx。
6. 德拉戈·科斯:"可持续发展:实现乌托邦?",《社会学杂志》第 54 卷,2012 年第 1 期,第 20 页。
7. 德雷克斯哈格和墨菲:《可持续发展》,第 19—20 页。
8. 格罗·哈莱姆·布伦特兰:"ARA 讲座",维也纳科技大学,2013 年 11 月 18 日,登录日期 2016 年 2 月 1 日,https://www.youtube.com/watch?v=X7Z2o8tZZoE。
9. 布伦特兰:"ARA 讲座"。

12 未来展望

要点

- 《我们共同的未来》中提出的可持续发展的定义,很有可能会在可预见的未来继续使用。
- 对基于环境、人类发展和经济增长三大支柱的可持续性的理解也可能扩展为多维模型。
- 《我们共同的未来》是一个受到高度赞扬的开创性文本,提供了最广泛使用的"可持续发展"的定义。

潜力

人类可能会继续面临《我们共同的未来》中提到的挑战,例如气候变化、人口过剩和自然资源减少。可持续发展仍将是国际政治议程中的首要议题,决策者很可能会继续借鉴布伦特兰所提供的关于可持续发展的经典定义。

报告中讨论的问题直至 21 世纪依然存在,许多问题更加严重。预计到 2050 年世界人口将达到 91 亿,人口增长主要发生在发展中国家。[1] 为了养活这些人口并避免粮食短缺,粮食产量将必须增长 70%。[2] 我们很难在粮食需求增加与自然资源(如肥沃的土地和水)迅速减少之间取得平衡。实现这种平衡需要资源管理上的重大改变和强烈的政治意愿。《我们共同的未来》提供的解决办法,如技术发展和国际合作,必须付诸实施。

如果找不到正确的解决方案,可能会导致严重的全球危机,威胁到人类的生存。正如南非宗教领袖和活动家德斯蒙德·图图所说:"我们只有一个世界,如果我们摧毁了它,我们就完了。"[3]

> "这是一个引人关注的现实,不能轻易被忽视。由于我们没有找到解决这些根本而严重问题的答案,除了继续努力寻找,我们别无选择。"
>
> —— 世界环境与发展委员会:《我们共同的未来》

未来方向

《我们共同的未来》提出了一种管理自然资源的革新方法,将环境与人类发展相结合,并把经济增长作为第三要素。这些是构成可持续发展概念的三大支柱,尽管人们对于这些支柱具体含义的理解尚存争议,但它们有助于我们更好地认识人类发展的复杂性。

为了解决这个问题,一些政治经济学家和社会科学家,包括哲学家,已经开始提供可持续发展的多维模型。例如,哥伦比亚裔美国人类学家阿图罗·埃斯科瓦尔提出了另外一种发展概念,"将融自然、文化和政治为一体的地方模式考虑在内"。[4] 文化是在最初的可持续发展三要素模型中加入的第四个要素。

另一个可以添加到该模型中的维度是全球治理——政府间政治合作的方式。显然,即使从《我们的共同未来》提供的例子来看,除非有具体的机制去落实,否则关于人类未来的创新思想将仅仅是空想。该报告预测,各国政府将不愿意将政府政策转向可持续性,并强调民间团体——一般公众——的游说对实现这一目标的重要性。这一预言合情合理。然而,当前人们对于民间团体在这一过程中如何发挥最大作用的认识尚未达成共识,因此,这一问题将会继续激发理论创新,并引发争论。

小结

《我们共同的未来》是联合国世界环境与发展委员会在 1987 年撰写的一份报告，也被称为《布伦特兰报告》，是以委员会主席、挪威前首相格罗·哈莱姆·布伦特兰命名的。该报告被公认为是可持续发展概念的权威文本。虽然也有一些早期作品提到了类似的概念，但正是《我们共同的未来》普及了人类发展和自然资源之间相互联系的理念，倡导负责地管理自然资源，以便为子孙后代维护好自然资源。可持续发展这一定义如今已得到公认，并常用于学术研究和政治语境中。

《我们共同的未来》介绍了当代发展政策的概念基石：经济增长、社会发展、自然和生态保护这三大支柱。该报告发表后立即受到了广泛的认可，并被牛津大学出版社誉为"十年中有关世界未来最重要的文件"。实际上，说该报告是 20 世纪最具影响力的文本之一也不为过。由于其中所讨论的挑战和可能的解决方案仍然具有现实性，因此该报告将继续影响 21 世纪的政治思想。

包括联合国在内的大多数国际组织一如既往地将可持续发展纳入其议程。联合国最近制定并通过了可持续发展目标，更新了 2000 年在纽约通过的《千年发展目标》，这是可持续发展在 21 世纪仍然具有重要意义的又一例证。毫无疑问，《我们共同的未来》在未来几十年仍会极具影响力。

1. 粮食及农业组织:"2050年如何养活世界",罗马:粮食及农业组织,2009年,第2页。
2. 粮食及农业组织:"如何养活世界",第2页。
3. 德斯蒙德·图图:"可持续发展是我们无法承受的奢侈品吗?",长老访谈,开普敦,2012年5月12日,登录日期2016年2月1日,https://www.youtube.com/watch?v=ArHey8SVJQE。
4. 阿图罗·埃斯科瓦尔:"地方文化:对全球主义和本土化底层战略的反思",《政治地理学》第20卷,2001年,第139—174页。

术语表

1. **21世纪议程**：1992年在里约热内卢举行联合国环境与发展会议，会议制定了在国际和国家层面实施可持续发展原则的行动计划。这是一项自愿协议。虽然得到欧洲国家强烈支持，但遭美国反对（这解释了为什么该计划没有充分实施）。

2. **博帕尔灾难**：1984年12月印度博帕尔市的一家农药工厂发生气体泄漏，这被认为是有史以来最严重的工业灾难。事故发生地中央邦地区至少有3 787人死亡。

3. **勃兰特报告**：由国际发展问题独立委员会起草的关于全球经济发展的报告，该委员会由西德政治领袖维利·勃兰特担任主席。

4. **布伦特兰委员会**：正式名称为世界环境与发展委员会，这是一个由联合国召集的机构，旨在促进各国之间的合作，以实现可持续发展。

5. **因果分析**：研究结果产生原因的一种方法。

6. **切尔诺贝利事故**：1986年4月在当时苏联，今天乌克兰境内一座核电站发生的一起工业事故。这是世界上最严重的核电站事故，影响了数千人。

7. **民间团体**：社会中除国家和商业部门以外的所有部分。

8. **气候变化**：由自然和人为原因造成的全球气候变化模式，与全球变暖有关。

9. **罗马俱乐部**：1968年在罗马林琴科学院成立的国际智库。其成员包括国际官员、政治领袖和杰出商人。俱乐部积极参与有关国际发展的辩论。

10. **冷战（1947—1991）**：美国与苏联及其各自盟国之间的政治紧张时期。虽然两国从未发生过直接的军事冲突，但两国进行了秘密的代理人战争，并相互进行间谍活动。

11. **企业社会责任（CSR）**：基于道德原则的企业自我监管体系。它促进了包括人、地球和利润的"三重底线"原则。

12. **古巴导弹危机**：1962年10月由于在古巴安装核武器，美国与苏联之间发生对抗。这是世界最接近核战争的时刻。

13. **发展**：旨在增加人们生活选择的政治和社会进程；主要是指政府和其他机构用来帮助、管理和规范社会的行为。

14. **地球峰会**：1992年6月在里约热内卢举行的联合国会议，推动了《关于环境和发展的里约热内卢宣言》和《21世纪议程》。

15. **经济及社会理事会（ECOSOC）**：联合国系统内负责经济和社会活动的主要机构之一。

16. **生态系统**：指包含存在于特定物理环境中所有有机体的生物系统。

17. **长老会**：2007年由南非政治家纳尔逊·曼德拉组建的一支由世界退休领袖组成的独立组织，他们为和平与人权而共同努力。

18. **环境经济学**：研究经济问题的一门学科，认为经济问题受到可持续性以及环保政策的影响。

19. **粮食短缺**：粮食供应不足导致营养不良、饥饿和饥荒。这种情况经常出现在发展中国家。

20. **联合国大会**：联合国的主要机构之一，也是唯一一家所有成员国都有平等代表权的机构。

21. **全球北方（发达国家）**：由北美和西欧组成的全球社会经济和政治类别。

22. **全球南方（发展中国家）**：由非洲、南美洲、亚洲和中东组成的全球社会经济和政治类别。

23. **漂绿**：假装环保却不采取任何真正环保措施的营销形式。

24. **人力资本**：对人类繁荣有价值的知识、技术、潜在创新和人类技能。

25. **国际可持续发展研究所**：一个独立的、非盈利性的组织，旨在通过可持续的方式促进人类发展。该组织报道国际谈判，鼓励创新和沟通，并尝试让公民、企业和政策制定者共同参与可持续发展进程。

26. **《京都议定书》**：1997 年在日本京都通过的一项国际条约，要求减少二氧化碳（CO_2）排放和减缓气候变化。

27. **千年发展目标**：联合国千年首脑会议于 2000 年制定的八项国际发展目标，原定于 2015 年实现。这些目标包括消除贫穷、普及初等教育、保护妇女权利、降低儿童死亡率、改善产妇健康、防治艾滋病毒/艾滋病等疾病、维护环境可持续性，并在全球范围内促进国际合作。

28. **监测和评估**：项目管理中不断检查某项活动以确定其有效性，并不断更新相关信息的过程。

29. **多边（主义）**：迈尔斯·卡勒提出的国际关系概念，意在促进多个国家的国际治理。

30. **自然资本**：自然资源的供应，包括地质、空气、水、土壤和生物体，可用于经济活动。

31. **新自由主义**：19 世纪出现的一种经济思想流派，提倡经济自由化和自由贸易政策。

32. **非政府组织（NGO）**：地方、国家或国际层面上的非营利性志愿团体。

33. **挪威劳工党**：1887 年成立的社会民主党，仍然活跃在挪威政治舞台上，提倡社会主义价值观。

34. **帕尔梅委员会**：裁军与安全委员会，由瑞典政治家奥拉夫·帕尔梅担任主席，1982 年发表了报告《共同安全》。

35. **巴黎会议**：2015 年 12 月在巴黎举行的联合国气候变化大会；会议通过了一项减缓气候变化的全球协议（巴黎协议）。

36. **政策**：政府、政党或公司的行动方针。

37. **"里约+20"峰会**：2012年在里约热内卢举行的联合国可持续发展大会，以跟进1992年地球峰会提出的行动。

38. **斯德哥尔摩会议**：也称联合国人类环境会议，于1972年6月5日至16日在斯德哥尔摩举行。该会议是制定可持续发展概念的重要一步，因为它拟订了平衡环境挑战和社会经济发展的行动原则，即《斯德哥尔摩宣言》。会议还提出了设立联合国环境规划署（UNEP）的想法。

39. **苏联**：苏维埃社会主义共和国联盟（苏联）是社会主义国家，存在于1922年至1991年间，以俄罗斯为中心。这是一个由共产党领导的一党制国家，首都在莫斯科。

40. **特使**：联合国秘书长选定的在人权或气候变化等问题上担任代表和顾问的人。这只是一个荣誉职位，没有报酬。

41. **强可持续性**：生态环保主义者支持该观点，认为人力资本和自然资本可能是互补的，但它们不可互换。

42. **可持续性**：在《我们共同的未来》中提出的负责地使用和管理资源的主要原则：一种"既满足当前需要又不损害后代满足自身需要的能力"的发展模式。

43. **可持续发展**：旨在改善人们生活同时确保后代能够满足其自身需求的各种政策和战略。

44. **可持续发展目标（SDGs）**：改善全球生活条件的17个目标，在2012年"里约+20"峰会上讨论了该观点，并于2014年被联合国大会接受。

45. **智库**：就特定的政治或经济问题提供建议和意见的一组专家。

46. **联合国（UN）**：由51个创始成员国于1945年创立的一个政府间组织，旨在促进国际合作与全球和平。

47. **联合国环境规划署（UNEP）**：联合国专门从事环境活动的机构。它是在莫里斯·斯特朗的倡议下于1972年创建的。总部设在肯尼亚的内罗毕。

48. **弱可持续性**：环境经济学支持的观点，认为人力资本和自然资本是可以互换的。

49. **世界环境与发展委员会**：布伦特兰委员会最初的名称，由联合国召集，目的是促进各国之间的合作，以实现可持续发展。

50. **世界卫生组织（WHO）**：联合国专门从事全球公共卫生的机构之一。总部设在日内瓦。

51. **南斯拉夫**：1918—1991年欧洲的一个国家。其领土包括今天的波斯尼亚和黑塞哥维那、克罗地亚、马其顿、黑山、斯洛文尼亚和塞尔维亚。

人名表

1. 吕西安·布沙尔（1938年生），加拿大政府的前环境部长，政治家和外交官，1996年至2001年期间，任加拿大魁北克地区的总理。

2. 布特罗斯·布特罗斯-加利（1922—2016），埃及政治家，1992年至1996年担任联合国秘书长。

3. 威利·勃兰特（1913—1992），1969年至1974年任德意志联邦共和国总理。1980年，他主持了一个独立委员会，编制了关于全球经济发展的《勃兰特报告》。

4. 莱斯特·R.布朗（1934年生），美国环保人士，写过有关当代世界主要挑战以及人类未来不同情景的书。他是智库世界观察研究所的创始人，独撰或者与别人合著了50多本书，其中包括《人类，土地和食物》(1963)。

5. 希拉里·克林顿（1947年生），美国政治家，曾在巴拉克·奥巴马总统任期内担任国务卿。2016年美国总统候选人。

6. 赫尔曼·达利（1938年生），马里兰大学公共政策学院的美国经济学教授，专门从事环境经济学研究。

7. 雅克·德洛尔（1925年生），法国经济学家，1985年至1995年担任欧盟委员会主席。

8. 尼丁·德赛（1941年生），印度经济学家，布伦特兰委员会成员，《我们共同的未来》作者之一。他还于1992年至2003年担任联合国主管经济和社会事务的副秘书长。

9. 阿图罗·埃斯科瓦尔（1952年生），哥伦比亚裔美国人类学家，研究社会运动和国际发展，是北卡罗来纳大学教堂山分校的人类学凯南特聘教授。

10. 马塞尼斯·芬格（1955年生），瑞士洛桑联邦理工学院技术管理学院继续教育学院的院长。

11. 沃尔克·豪夫（1940年生），德国政治家，社会民主党成员。

12. 韦斯·杰克逊（1936年生），可持续农业运动的创始人和领导人之一。研究人类农业活动与生态系统之间的关系，并呼吁要更谨慎地使用技术和不可再生的资源。

13. 潘基文（1944年生），韩国前外交和贸易部长，联合国前秘书长。

14. 德拉戈·科斯（1961年生），斯洛文尼亚律师、记者和前警官，经合组织反腐工作组的首任主席。

15. 托马斯·罗伯特·马尔萨斯（1766—1834），英国教士和学者，讨论了饥荒和人口增长之间的关系，也就是所谓的"马尔萨斯灾难"。1798年，他出版了他最重要的著作《人口论》。

16. 纳尔逊·罗里拉拉·曼德拉（1918—2013），南非人权活动家，1994年至1999年担任南非总统。

17. 弗朗索瓦·莫里斯·阿德里安·玛丽·密特朗（1916—1996），法国社会党领袖，1981年至1995年担任法国总统。

18. 大来佐武郎（1914—1993），日本经济学家和外交部长（1979—1980），负责研究日本经济增长问题。

19. 斯文·奥拉夫·约阿希姆·帕尔梅（1927—1986），瑞典首相。曾担任裁军与安全委员会（又称帕尔梅委员会）主席，于1982年出版报告《共同安全》。

20. 哈维尔·佩雷斯·德奎利亚尔（1920—2020），秘鲁前总理、联合国第五任秘书长，他委托格罗·布伦特兰担任世界环境与发展委员会主席，撰写《我们共同的未来》（即《布伦特兰报告》）。

21. 迈克尔·E.波特（1947年生），美国经济学家，著有关于企业社会责任和绿色商业的文章。哈佛商学院战略与竞争力研究所的威廉·劳伦斯主教大学教授。提出了"五力模型"，用以分析行业和商业发展竞争水平的框架。

22. 杰弗里·D.萨克斯（1954年生），美国经济学家、发展与贫困问题专家，也是联合国秘书长关于可持续发展的特别顾问。著有《贫穷的终结》（2005）、《共同财富》（2008）和《文明的代价》（2011）。

23. **范达娜·席瓦**（1952年生），印度环保人士、女权主义者和国际活动家。领导反对基因工程的运动，并与发展中国家的基层组织密切合作。

24. **莫里斯·斯特朗**（1929—2015），加拿大商人、联合国环境规划署（UNEP）首任署长，《我们共同的未来》作者之一。

25. **德斯蒙德·姆皮洛·图图**（1931年生），退休的主教和人权活动家，在南非与种族歧视作斗争。

26. **希夫·维斯瓦纳坦**，印度公共知识分子和社会科学家。

27. **鲍里斯·叶利钦**（1931—2007），俄罗斯联邦第一任总统。

WAYS IN TO THE TEXT

KEY POINTS

- Born in 1939, Gro Brundtland was the first female prime minster of Norway and is a prominent figure in the United Nations* (or UN, an international organization founded to foster cooperation between governments); in 1983 she was appointed as the chair of the UN World Commission on Environment and Development* (the Brundtland Commission).

- As a trained medical doctor and an experienced politician, Brundtland came up with a visionary approach to managing environmental and social resources, by understanding that environmental challenges and human development* (political and social processes aimed at increasing people's life choices) were fundamentally linked.

- *Our Common Future* (1987), also known as the *Brundtland Report*, is one of the main reference works on sustainable development* (the means by which societies can support themselves while preserving natural resources); it has provided the most cited definition of "sustainability" * in both academic and policymaking* circles.

Who Is Gro Brundtland?

Gro Harlem Brundtland was born in 1939 in Oslo, Norway. Her father, Gudmund Harlem, was a medical doctor and a prominent politician, like Brundtland. Speaking of her father's influence on her career choice, she says, "I'm proud of my father for voting Labor [the Norwegian Labor Party*: a social-democratic political party] because it's right, even though he does not benefit from it." ¹ Brundtland studied medicine at the Harvard School of Public Health in the United States, where she was exposed to "the seeds of internationalism" [2]. ("Internationalism" here refers to a political stance that favors cooperation between different states and peoples.)

Brundtland began work as a doctor, but in 1974 she put aside her medical career to become Norway's minister for environmental affairs. In 1981, she became prime minister of Norway, the first woman to hold this position. Brundtland recognized the connection between human health and the state of the environment. [3]Her background was unique in that she was the only prime minister who had ever held an environmental post. The then-secretary-general of the UN, Javier Perez de Cuellar, * appointed her to chair the World Commission on Environment and Development and to lead a "call for political action" on environment and development. [4]

Brundtland gathered an extraordinary group of commissioners, both academics and politicians. Together they came up with visionary ideas about human development and compiled their findings in a report under the title *Our Common Future*. This has since been referred to as "the most important document of the decade on the future of the world," according to Oxford University Press. [5]

What Does *Our Common Future* Say?

Our Common Future was a report commissioned to formulate "a global agenda for change" [6]and produced by the Brundtland Commission, * a body convened by the UN with the aim of fostering cooperation between nations to pursue sustainable development. It was conceived as a response to a growing number of serious concerns faced by the global community in the 1980s. The report recognized the need for long-term strategies for the management of natural resources. The Brundtland commissioners agreed that it was necessary to protect the environment, while at the same time stimulating economic and social development through international cooperation. Taken together, these goals were summarized as "sustainable development," * defined in the report as humanity's ability "to ensure that it meets the needs of the present without compromising the ability of future generations to meet their own needs." [7] This recognition of our responsibility to the future is captured in the title of the report.

Our Common Future has been translated into over 20 languages and was first distributed to politicians, together with accompanying educational video materials. [8] The report is an appeal to humanity to change its behavior in order to reverse dangerous environmental and social trends, both on an individual and a global level. In 1987, it was published as a book by Oxford University Press, a unique occurrence for a work that started life as a UN report. [9] By 1989, half a million copies had been sold. [10]

Our Common Future introduced the term "sustainability" into the everyday language of international politics. In the world of academia, discussion of sustainable development has led to the birth of a new discipline of environmental economics* (the study of economic matters as they are affected by issues of sustainability, and of the economic consequences of policies designed to protect the environment) and has also received attention from many other disciplines, such as human geography (the study of interactions between human populations and the environment), international relations (the study of the relations between and among nations and international institutions), and social anthropology (the study of the ways in which humans build and maintain social institutions such as laws and kinship). *Our Common Future* still serves as a reference book for both academic research and policymaking.

It is hard to overstate the importance and influence of *Our Common Future*. The points made in the report have become generally accepted, and Gro Brundtland's expert advice has been sought by a number of UN agencies, including the General Assembly* (the body of the UN in which all the world's nations are represented) and the World Health Organization* (or WHO, the agency dealing with global public health). The current secretary-general of the UN, Ban Ki-moon, * has appointed Brundtland one of his three special envoys* on environmental issues and climate change. * ("Special envoy" here refers to a person chosen by the secretary-general to act as a representative and adviser on a particular matter.)

Why Does *Our Common Future* Matter?

Our Common Future is widely acknowledged as the main source for the concept of sustainable development. It has introduced the term "sustainability" and the conceptual cornerstones of contemporary development policies—the three pillars of economic growth, social development (an improvement in matters such as living standards, healthcare, and institutions like the courts) and conservation of nature. While Brundtland and her commissioners were not the first to use this term, they were the first to give it such a wide international exposure and, as a result, they have made it a frequently used concept.

The report raised universal issues related to balancing human requirements with responsible management of the planet's natural resources. The book's insights into the need to stop the harmful overuse of natural resources were a revelation to many readers. [11] The authors were among the first in the field of international policy to connect environmental protection and economic development. Seeing the fulfillment of human needs as the major goal of development, [12] they mapped out the serious negative effects of damage to the environment, explaining how these threatened human well-being in the present and the future.

Brundtland and her commissioners cataloged and analyzed the major challenges faced by human societies at the time of publication; among these were poverty, deterioration of finite natural resources, and climate change. They then discussed possible solutions on a global and national scale, such as better international cooperation and raising awareness about the environment among politicians and the general public. The writers recognized the global scale of the challenges discussed in the report, believing that changes in international policy and increased awareness were the key to sustainability.

Having established the main conceptual framework of sustainability, the commissioners discussed the future and made what seems in retrospect an overly optimistic estimation that humanity would complete the global shift into sustainability by the year 2000. The world community still plans to reach Sustainable Development Goals (SDGs)* by 2030, though this is a full 30 years later than the original schedule proposed by Brundtland.

The Brundtland commissioners also identified the organizations and people who might bring about this change. They predicted that national governments would be resistant to the radical changes they proposed. As a result, the main push for policy change was expected to come from civil society* groups—political groups made up of ordinary people rather than professional politicians—who would work with both the general public and policymakers.

---

1. Gro Harlem Brundtland, *Madam Prime Minister: A Life in Power and Politics* (New York: Farrar, Straus and Giroux, 2005), 19.
2. United Nations, *Biography of Dr Gro Harlem Brundtland* (Geneva: United Nations, 2014), 1.
3. World Health Organization, "Dr Gro Harlem Brundtland, Director-General," Geneva: WHO, 1998, accessed March 9, 2016, http://www.who.int/dg/brundtland/bruntland/en/.
4. World Commission on Environment and Development (WCED), *Our Common Future* (Oxford: Oxford University Press, 1989), x.
5. WCED, *Our Common Future*, publisher's note, cover page.
6. WCED, *Our Common Future*, viii.
7. WCED, *Our Common Future*, 8.
8. Pratap Chatterjee and Matthias Finger, *The Earth Brokers: Power, Politics and World Development* (London: Routledge, 2013), 84.
9. WCED, Report of the World Commission and Development on Environment and Development: Our Common Future, August 4, 1987, accessed February 1, 2016, http://www.un-documents.net/wced-ocf.htm.
10. Linda Starke, *Signs of Hope: Working towards Our Common Future* (Oxford: Oxford University Press, 1990), 3.
11. Winin Pereira and Jeremy Seabrook, *Asking the Earth: Farms, Forests and Survival in India* (Sterling, VA: Earthscan, 1990), 62.
12. WCED, *Our Common Future*, 44.

SECTION 1
INFLUENCES

MODULE 1
THE AUTHOR AND THE HISTORICAL CONTEXT

KEY POINTS
- *Our Common Future* (the *Brundtland Report*) provided the classic definition of sustainable development* still used today.
- The report inspired an important shift in international political thinking, raising awareness about the urgent need to manage natural resources in a responsible manner so that they would be available for future generations.
- Major challenges named in the report, such as climate change, * are still relevant today.

Why Read This Text?

Gro Brundtland's *Our Common Future* (1987) is the key text for the concept of "sustainable development" : human society's ability to ensure that the various economic and institutional improvements necessary to increase a population's life choices "meets the needs of the present without compromising the ability of future generations to meet their own needs." [1] The text's publication marked the beginning of an important period in our history, as it was the first time that high-level politicians recognized that the business-as-usual approach to managing the world's natural resources would eventually lead to global environmental collapse. The authors called for a total rethink of production and distribution at a global level.

The report gave the first comprehensive outline of major issues such as a fast-growing world population, food insecurity* (lack of access to food), increasing urban populations, environmental degradation, and climate change. By March 1989, 22 national governments had created plans for achieving sustainability, * demonstrating their acceptance of the report and of the new concept of sustainable development. ² The report continues to influence policymaking* processes today.

> *"I decided to accept the challenge. The challenge of facing the future, and of safeguarding the interests of coming generations. For it was abundantly clear: We needed a mandate for change."*
>
> —— Gro Brundtland, Chairman's Foreword,
> *Our Common Future*

Author's Life

Gro Brundtland was born in 1939 in the Norwegian capital, Oslo. She was trained as a medical doctor, following in the footsteps of her father, who was a doctor and a prominent Norwegian politician. She went on to study at the Harvard School of Public Health, where she refined her views on health and human development. * From an early age she had been involved in political activism, her father having enrolled her as a member of the left-wing Norwegian Labor* Movement when she was just eight years old.

On returning to Norway in 1965 after completing her graduate studies at Harvard, Brundtland joined the Norwegian Ministry of Health, where she worked as a medical consultant. In 1974, she accepted a nomination to become minister for environmental affairs and in 1981 she became the youngest person and the first woman to serve as prime minister of Norway. She stayed in office for two terms (1986–9 and 1990–6). Her unique background as a minister for the environment and a prime minister later led to her being chosen to lead the World Commission on Environment and Development*—a body instituted by the United Nations* to find solutions to the pressing problem of development in a world of finite resources.

Brundtland assembled an international team of leaders to author *Our Common Future*, consisting of 21 prominent academics, politicians, and officials from different nations. Brundtland described the work of her commission as in "the spirit of friendship and open communication," reflecting the learning process of people with "different views and perspectives, different values and beliefs, and very different experiences and insights." Despite this diversity, the report's writers arrived at a shared conclusion about the urgent need for sustainability. [3]

Author's Background

Gro Brundtland followed her father in becoming a prominent Norwegian politician and a practicing medical doctor. Before she was nominated to lead the writing of *Our Common Future*, she had participated in other UN commissions that dealt with questions of disarmament (the decommissioning of weapons of war), poverty, and the environment. As an experienced and internationally recognized politician, she was familiar with the main issues of the time.

Our Common Future can be considered a product of its era. It reflects the urgency of the task of formulating sustainable development strategies in "a compelling reality" of political, economic and environmental problems. [4] The Cold War* (1947–91), a period of grave tension between the United States and its allies and the now-dissolved Soviet Union* and its allies, was still ongoing at the time. The world community faced a proliferation of massive public (that is, state) debt in developing countries, reduced aid payments from richer to poorer countries, and growing poverty. [5] Famines in Africa, the Indian Bhopal disaster* of 1984 (the worst industrial disaster in history) and the Russian Chernobyl accident* of 1986 (the worst nuclear power-plant disaster to date) are mentioned in the report as examples of how the future of humanity could be endangered unless changes in attitudes and policies were made. [6]

Many of the challenges mentioned in the report still remain important today, such as climate change, food insecurity, and environmental degradation. This explains why Gro Brundtland is still actively involved in international politics as a special envoy to the current UN secretary-general, Ban Ki-moon* of Korea, on climate change.

---

1. World Commission on Environment and Development (WCED), *Our Common Future* (Oxford: Oxford University Press, 1989), 8.
2. Linda Starke, *Signs of Hope: Working towards Our Common Future* (Oxford: Oxford University Press, 1990), 3.

3. WCED, *Our Common Future*, xiii.
4. WCED, *Our Common Future*, xix.
5. WCED, *Our Common Future*, x.
6. WCED, *Our Common Future*, 3.

MODULE 2
ACADEMIC CONTEXT

KEY POINTS

- *Our Common Future* raised the important question of how human society can prosper despite decreasing natural resources.
- Disagreeing with the assumption that a growing human population would have catastrophic consequences, the members of the Brundtland Commission* remained optimistic that a new concept of sustainability*— the requirement that we live within our finite natural resources—could address many of these concerns.
- The report presented the hopes and ideals of its authors and their contemporaries about a better future for humanity.

The Work in Its Context

Gro Brundtland's report *Our Common Future* (1987) has been recognized for its contribution to the evolution of the concept of sustainable development—the capacity to increase living standards while ensuring that future generations will not find themselves without resources in an environmentally degraded world. The report recognized the concepts of "development" * and "environment" as inseparable, since it was becoming clear that failures of development policies and mismanagement of natural resources were deeply interconnected. [1]

The concept of sustainability first came to public notice in the 1970s, in the American environmentalist Wes Jackson's* work on agriculture. Jackson argued that nonrenewable resources should be used in the most efficient possible way. ²

The question of how humans can survive even though the planet's available resources are finite was discussed before *Our Common Future*. For example, in 1972 the Club of Rome* (an international think tank* that united academics, businessmen, and politicians) used computer modeling to produce the influential report *The Limits to Growth*. ³ The report offered statistical evidence identifying future trends of overpopulation and growing demand for natural resources; the authors warned that these would be irreversible unless a better, more sustainable, approach to the management of natural resources was put in place.

> "The challenge of finding sustainable development paths ought to provide the impetus—indeed the imperative—for a renewed search for multilateral solutions and a restructured international economic system of co-operation. These challenges cut across the divides of national sovereignty, of limited strategies for economic gain, and of separated disciplines of science."
>
> —— World Commission on Environment and Development,*
> *Our Common Future*

Overview of the Field

The relationship between finite environmental resources and human survival has been widely discussed since the work of the eighteenth-century English demographer Thomas Malthus* ("demography" refers to the statistical study of the makeup of a community or population). His *An Essay on the Principle of Population* (1798) argued that there is a link between population growth, food availability, and a limited "power in the earth to produce subsistence for man." [4] Malthus introduced a hypothetical scenario in which a world population could outreach the food supply available, and argued that population growth should be limited—similar issues to those identified in the Club of Rome's report.

In the period just before the publication of *Our Common Future* it was already becoming clear that the earth's resources had to be managed sustainably. This idea challenged the theory, then dominant in mainstream economics, that the problems presented by finite environmental resources and environmental degradation could be solved by the relocation of people from one region of the world to another. [5]

Academic Influences

There were earlier attempts to develop the concept of ecological sustainability at the United Nations* Conference on the Human Environment held in Sweden in 1972, also known as the Stockholm Conference. * This was the first global conference to evaluate the environmental consequences of human development at a macro (that is, general and global) level and to elaborate universal guidelines to protect and improve the environment. [6] The conference passed a Declaration on Human Development, which acknowledged that every individual has a right to a healthy environment. National governments were given responsibility for implementing and safeguarding policies to protect the natural environment and the UN officially recognized the importance of addressing the world's environmental problems. [7]

In the same year, the Club of Rome issued its report *The Limits to*

Growth, identifying future trends such as overpopulation and a growing demand for natural resources. The report was well received, and many politicians were immediately convinced by its argument about the correlation between overpopulation and the devastation of natural resources. Later on, however, the Club of Rome was criticized for using incorrect data. [8] One of the Brundtland commissioners, Japanese economist Saburo Okita, * was also a member of the Club of Rome.

The Brundtland commissioners were aware of the ongoing academic and political debates around global environmental problems. They were the experts in the field at the time, and called for more international cooperation. The commissioners argued, for example, that countries of the Global North* (richer countries, mostly in the northern hemisphere) should offer more consistent support to their poorer counterparts in the Global South* (impoverished countries, mostly in the southern hemisphere).

---

1. World Commission on Environment and Development, *Our Common Future* (Oxford: Oxford University Press, 1989), 30.
2. Wes Jackson, *New Roots for Agriculture* (Lincoln: University of Nebraska Press, 1985), 144.
3. See Donella H. Meadows et al., *The Limits to Growth: A Report for the Club of Rome's Project on the Predicament of Mankind* (New York: Universe Books, 1974).
4. Thomas Robert Malthus, *An Essay on the Principle of Population* (London: J. Johnson, 1798), 13.
5. Jorgen Norgard et al., "The History of the Limits to Growth," *Solutions* 2, no. 1 (2010): 59–63.
6. Felix Dodds et al., *Only One Earth: The Long Road via Rio to Sustainable Development* (London: Routledge, 2012), 8–11.
7. Dodds et al., *Only One Earth*, 11–12.
8. Graham Turner, *A Comparison of Limits to Growth with Thirty Years of Reality*, CSIRO Working Paper (Canberra: CSIRO, 2008), 36.

MODULE 3
THE PROBLEM

KEY POINTS

- The core question of *Our Common Future* was how to ensure the economic growth necessary for social development* while using and managing finite natural resources in a responsible way.
- The topics addressed in the report had previously been addressed by other contemporary scholars, including the think tank* the Club of Rome, * the American environmentalist Lester Brown, * and those engaged in the new field of environmental economics. *
- The authors of the report were criticized for prioritizing economic growth over the protection of natural resources; only one of them directly responded to this criticism.

Core Question

In *Our Common Future*, Gro Brundtland addressed four strategic goals:

- "To propose long-term environmental strategies for achieving sustainable development* by the year 2000."
- "To achieve greater cooperation among developed and developing countries."
- "To optimize strategies for addressing environment concerns."
- "To help define shared perceptions of long-term environmental issues." [1]

The core question of the report was how to secure the survival and well-being of humankind in the future through safeguarding the environment. [2] The Brundtland commissioners* approached this question through an understanding that the planet's existing ecological system could not support the growing needs of humans without a major shift in international policies related to natural-resource management. The imbalance between limited environmental resources and growing human needs was illustrated by problems such as economic crises, deepening global poverty, and life-threatening accidents, both natural and man-made. For example, major industrial accidents in India at a pesticide factory in Bhopa* and in what is today Ukraine at the Chernobyl* nuclear power-plant, both during the 1980s, led to deaths, injuries, and long-term negative environmental effects. These events forced the global community to reassess the danger posed by technogenic catastrophes (accidents caused by technology). Better international cooperation was discussed as a way to prevent this kind of incident. [3]

While no early drafts of the report have been published, Brundtland commissioners have revealed that the initial title of the report was *A Threatened Future*. [4] The commissioners were dedicated to defending humanity against future environmental and social crises.

> "The environment does not exist as a sphere separate from human actions, ambitions, and needs, and attempts to defend it in isolation from human concerns have given the very word 'environment' a connotation of naivety in some political circles."
> ——World Commission on Environment and Development,*
> *Our Common Future*

The Participants

An international team, known as a "commission," wrote the report under Brundtland's leadership. It was composed of 21 prominent academics, politicians, and international officials from Algeria, Brazil, Canada, China, Colombia, Cote d'Ivoire, Germany, Guyana, Hungary, India, Indonesia, Italy, Japan, Nigeria, Norway, Saudi Arabia, the Soviet Union, * Sudan, the United States, Yugoslavia, * and Zimbabwe. Among them were Maurice Strong, * a Canadian businessman and the first president of the United Nations Environment Program (UNEP), * Nitin Desai, * an Indian economist, and Saburo Okita, * a former Japanese minister and member of the Club of Rome, who led Japan to its postwar economic growth. The 21 participants who produced the report are referred to as "commissioners."

The commission was an independent international body and aimed to provide a full analysis based on numerous qualitative and quantitative data sets—a mixture of numerical information and other kinds of research, such as descriptions and interviews. To ensure that the most up-to-date information was used, the commission held five public hearings with indigenous people, farmers, and scientists. The authors described *Our Common Future* as written collaboratively by many people in "all walks of life." [5]

There had been previous attempts to provide a comprehensive understanding of how to manage earth's diminishing resources, such as *The Limits to Growth* by the Club of Rome, and articles and books by the American environmentalist Lester Brown. [6] Both the Club of Rome and Brown considered a number of possible scenarios for managing population growth and finite natural resources, sometimes even predicting apocalyptic outcomes in which competition for access to natural resources would bring about a final global catastrophe. In contrast, the Brundtland commissioners had faith in humanity's ability to solve these problems peacefully by means of sustainable development. With the exception of one commissioner, the Indian economist Nitin Desai, who eventually rejected the concept, every member of the team that produced *Our Common Future* continued to use the concept of sustainable development in his or her later work.

The Contemporary Debate

Although the report introduced the concept of sustainable development as we know it today, it did not explain how sustainability could be achieved through the management of natural resources. It adopted an anthropocentric approach (one centering on human needs), rather than a nature-centric, or ecocentric, approach (one emphasizing nature above human needs). [7] Overall, *Our Common Future* supports an economic argument known as "weak sustainability,"* an approach to the management of natural capital* (natural resources beneficial to human prosperity) that sees it as potentially interchangeable with or replaceable by human capital* (knowledge, technology, and innovation).

For this reason, the report has been heavily criticized by ecological economists, who argue that it falsely offers "a sense of comfort to the effect that the environment can be dispensed with," [8] and implies incorrectly that implementing sustainability will not require much radical change. Brundtland commissioners addressed this critique in different ways. One member of the team, Nitin Desai, changed his views about sustainability in response to criticisms. He later voiced opposition to some of the content of the report he had helped to produce. [9]

In contrast, another Brundtland commissioner, Canadian businessman and environmentalist Maurice Strong, remained committed to the views expressed in the report. Strong understood sustainable development as a set of policies arranged "on the premise that population and per capita consumption [consumption per head of population] must operate within the global ecosystem* to respond indefinitely to our demand on resources and to assimilate the wastes produced." [10] (An "ecosystem" is a biological system comprising all the organisms that exist in a specific physical environment.)

---

1. World Commission on Environment and Development (WCED), *Our Common Future* (Oxford: Oxford University Press, 1989), viii.
2. WCED, *Our Common Future*, xi.
3. WCED, *Our Common Future*, 95–235.
4. Linda Starke, *Signs of Hope: Working towards Our Common Future* (Oxford: Oxford University Press, 1990), 1.
5. WCED, *Our Common Future*, xiii.
6. Lester R. Brown, *Man, Land and Food: Looking Ahead at World Food Needs* (Washington DC: US Department of Agriculture, 1963).
7. Parenivel Mauree, "Sustainability and Sustainable Development," *Le Mauricien*, August 17, 2011, accessed February 1, 2016, http://www.lemauricien.com/article/maurice-ile-durable-sustainability-

and-sustainable-development.
8. P. A. Victor et al., "How Strong is Weak Sustainability?" *Economie Appliquée* 48, no. 2 (1995): 75–94.
9. Nitin Desai, "Symposium: The Road from Johannesburg," keynote address, (Georgetown: *Environmental Law Review*, 2003).
10. Maurice Strong, *Where on Earth We are Going?* (Toronto: Vintage Canada, 2001), 195.

MODULE 4
THE AUTHOR'S CONTRIBUTION

KEY POINTS

* Gro Brundtland was determined to articulate the way in which the environment and development* are interconnected, and to emphasize the need to manage natural resources in a responsible manner.
* At the time of publication of *Our Common Future*, the idea of sustainability* was an almost revolutionary way of thinking about managing economies in that it took natural resources into account.
* Brundtland's previous involvement in international commissions gave her extensive knowledge of progressive ideas about international development.

Author's Aims

Gro Brundtland's work on *Our Common Future* began in December 1983 when the secretary-general of the United Nations, * Javier Perez de Cuellar, * appointed her chair of the World Commission on Environment and Development. Then the Norwegian prime minister, she saw this appointment as "the challenge of facing the future, and of safeguarding the interests of coming generations" [1] and as an opportunity to work toward the ideals she had developed as a participant in two previous UN commissions, the *Brandt Report** on global economic development and the Palme Commission* on disarmament and global security. In her role as the chair of the Palme Commission, she aimed to achieve two main interconnected goals: to persuade political leaders to return to multilateral* disarmament (that is, a policy implemented simultaneously by a number of countries working together), despite the threat of nuclear war that characterized the ongoing Cold War* between the United States and the Soviet Union* and their allies; and to expand the understanding of human development.

Although disappointed by the deterioration in global cooperation throughout the 1970s and 1980s, Brundtland remained hopeful that progressive ideas could be brought into reality as a result of a new awareness promoted at the international level. By raising concerns about instances of environmental degradation, she aimed to prove that these were shared human challenges that required a unified international response and a more equitable distribution of global financial resources. In *Our Common Future*, environmental concerns were presented as closely related to social concerns. For example, climate change* and soil depletion (infertility in soil arising from overuse by farmers) were recognized as complex problems threatening human survival that could not be solved by developing (poorer) nations on their own, but required aid from richer nations.

> "But the 'environment' is where we all live; and 'development' is what we all do in attempting to improve our lot within that abode. The two are inseparable. Further, development issues must be seen as crucial by the political leaders who feel that their countries have reached a plateau towards which other nations must strive."
> ——World Commission on Environment and Development,*
> *Our Common Future*

Approach

Environmental concerns had been discussed at previous UN conferences such as the Stockholm Conference* of 1972. These conferences provided precedents for the Brundtland Commission. The most important achievement of the commission was the recognition of the link between the state of the natural environment and human well-being. Brundtland aimed to develop these ideas further in *Our Common Future*. She challenged the initial terms of reference for her commission, insisting that its considerations should not be limited to environmental issues; instead, she argued that environmental concerns should be understood in relation to human development. *Our Common Future* stated that the environment and development were in fact inseparable. The major concept of the report—sustainable development*—was formulated on the basis of this idea.

According to the report, sustainable development has limits created by the current level of technological development, existing social organizations, and the planet's ability to renew its resources. The report also introduced the element of economic growth into the understanding of development. Sustainable development was presented as a three-part challenge comprising environmental protection, social development (the bettering of human life choices), and economic growth.

Contribution in Context

Our Common Future's expansion of the meaning of environmental concerns was novel. Brundtland and her colleagues argued that the environment should not be considered "as a sphere separate from human actions, ambitions, and needs." They added that "attempts to defend it in isolation from human concerns have given the very word 'environment' a connotation of naivety in some political circles." [2] For the first time, the UN took the official position that the environment was interconnected with human development and economic growth.

The report's mention of the needs of future generations was also new. Unlike most UN reports, designed as internal documents to be circulated only within UN agencies, *Our Common Future* was published as a book. This was a result of its novelty and the urgency of the shift toward multilateral international development suggested in the report. The report called on all the world's citizens to work together to bring about the sustainability required to ensure human survival.

---

1. World Commission on Environment and Development (WCED), *Our Common Future* (Oxford: Oxford University Press, 1989), ix.
2. WCED, *Our Common Future*, x.

SECTION 2
IDEAS

MODULE 5
MAIN IDEAS

KEY POINTS

- *Our Common Future* identified major challenges faced by humanity in the twentieth century, including environmental crises, poverty, the depletion of natural resources, climate change, * and political tensions such as the Cold War. *
- The report defines sustainability* as humanity's ability "to ensure that it meets the needs of the present without compromising the ability of future generations to meet their own needs."
- The report aimed to raise awareness among politicians and the general public about the need to change the pattern of overconsumption of natural resources.

Key Themes

The core question posed by Gro Brundtland's *Our Common Future* (1987) was how to secure the survival and well-being of humankind in the distant future while safeguarding the environment. [1] The Cuban missile crisis* of 1962—perhaps the closest the United States came to nuclear war with its Cold War adversary, the Soviet Union*—and a number of industrial accidents with serious environmental effects, such as the Indian Bhopal disaster* and the Soviet nuclear power-plant catastrophe at Chernobyl, * all had long-term negative effects on the natural environment and human health. Politicians and other Western leaders, including Brundtland and her commissioners, wanted to prevent such crises in the future and to enhance international cooperation.

The key theme of the report is the urgent need to move toward sustainable development. * This theme arises from the understanding that growing human needs did not fit within the existing ecological system of the planet and that a major review of international policies of natural resource management was essential.

> "Humanity's inability to fit its activities into that pattern is changing planetary systems, fundamentally. Many such changes are accompanied by life-threatening hazards. This new reality, from which there is no escape, must be recognized—and managed."
> —— World Commission on Environment and Development,*
> *Our Common Future*

Exploring the Ideas

Our Common Future aims to convince the reader that sustainable development is of the utmost importance. The report demanded serious attention from politicians and the general public, arguing that immediate action should be taken to review unsustainable practices in managing natural resources. [2] It drew attention to major issues in contemporary human society, such as the importance of raising public awareness about environmental concerns. These concerns are organized into a set of concrete problems, referred to as challenges, which result from unsustainable trends. Later in the report, possible solutions are introduced and discussed.

Following a short introduction written by the commission

chairperson Gro Brundtland that provides information about the commissioning of the report and the commissioners who worked on it, the main body of the report introduces the key argument immediately: "Hope for the future is conditional on decisive political action now to begin managing environmental resources to ensure both sustainable human progress and human survival." [3]

The text's opening overview introduces the global concerns and challenges of human development, * and then suggests possible solutions in the form of international cooperation and institutional reform. The global concerns presented in the main part of the report include the question of human survival in the future, the economic collapse triggered by poverty, crises, and economic decline, and the overexploitation of natural resources. [4] Global challenges, referred to as "common challenges," are named and discussed, such as the rapid growth of the human population, food insecurity, * climate change, the extinction of many species of flora and fauna, energy insecurity, the hazards of industrialization (the large-scale increase in industrial production), and urban development issues. [5]

The report's concluding discussion of possible solutions, called "common endeavors," suggests areas where international cooperation is needed, such as in the management of common global resources, like the oceans, outer space, and Antarctica, and the need for disarmament, clean energy, and institutional and legal actions to secure global peace and cooperation. [6]

Language and Expression

Although *Our Common Future* is a report commissioned by the

United Nations* General Assembly, * its structure is quite different to the traditional format of UN reports that usually give brief background information and concentrate on proposed actions. It is also unusual for a UN report to be published as a book, as *Our Common Future* was.

The authors designed the report to be accessible to the widest possible audience, and aimed "to translate [our] words into a language that can reach the minds and hearts of people young and old." [7] Since the report was targeted at everyone, its argument had to be developed and presented in a simple but powerful way. As a result, it appears to be more comprehensive than other reports on the subject. It elaborates its main topic and attempts to convince readers that sustainable development has immense significance.

Our Common Future demands serious attention from a global audience and calls for immediate action at every level of human activity. [8] It expresses major concerns about contemporary human society, such as climate change, food insecurity, pollution, and a fast-growing population. These concerns are presented as a set of concrete problems, referred to as "challenges," that result from unsustainable general trends. Possible solutions are also introduced and discussed.

---

1. World Commission on Environment and Development (WCED), *Our Common Future* (Oxford: Oxford University Press, 1989), xi.
2. WCED, *Our Common Future*, viii.
3. WCED, *Our Common Future*, 1.

4. WCED, *Our Common Future*, 27–36.
5. WCED, *Our Common Future*, 95–257.
6. WCED, *Our Common Future*, 261–343.
7. WCED, *Our Common Future*, xiv.
8. WCED, *Our Common Future*, x.

MODULE 6
SECONDARY IDEAS

KEY POINTS

- In its discussion of current challenges and possible solutions, *Our Common Future* used a "causal analysis" * to distinguish between the causes and consequences of the crises.
- This causal analysis was later developed into a new approach to policy* analysis, that of monitoring and evaluation*—a process in which an activity is constantly checked to see if it is effective, and information about it is continuously updated.
- Based on monitoring and evaluation, *Our Common Future* praises nongovernmental organizations (NGOs*) as a principle force in policy change. (NGOs are nonprofit organizations independent of the government of the nation in which they operate.)

Other Ideas

Applying causal analysis to policymaking by distinguishing between the causes and consequences of significant crises, as Gro Brundtland's *Our Common Future* did, was groundbreaking. Previously, it was mainly philosophers, not policymakers, who explained situations in terms of causes and effects. Calling for a deeper analysis, *Our Common Future* succeeded in influencing the approaches employed by other policymakers and NGOs in their management of environmental resources. The report inspired a great deal of research founded on monitoring and evaluation, according to which the researcher continuously assesses and tracks results. Since then, many economists, public managers, and biologists (those engaged in the study of living things) have developed methods to estimate the importance and impact of environmental and socioeconomic policies on ecosystems. *[1]

Brundtland and her commissioners believed in humankind's capacity to change behavioral patterns that negatively impact the environment, and to cooperate at all levels. [2] This belief, clearly conveyed in the report, can be partially explained by the moment in which the report was written and the aspirations of the generations that had lived through the Cold War* years of 1947 to 1991. It was a time of relative optimism and many people believed at that point that governments could work together to reshape the future.

Although the report also stimulated a change in the attitudes and political thinking of the era, the signs of change were often interpreted overly optimistically. As a result, complex issues were simplified into a common solution: cooperation between nations.

> "We must understand better the symptoms of stress that confront us, we must identify the causes, and we must design new approaches to managing environmental resources and to sustaining human development."
> ——World Commission on Environment and Development,*
> *Our Common Future*

Exploring the Ideas

Although *Our Common Future* identified a link between the environment and human development, * this link required further study to understand existing patterns, to find tools to measure them, and to elaborate necessary policy interventions (changes in policy intended to produce a certain effect). Imbalanced approaches risked causing new crises rather than solving existing ones.

The report had the potential to bring about comprehensive systematic changes, but did not provide the methodological tools—the specific, technical, methods—needed to help tackle actual crises. The double crisis of the natural environment and human development is complex, and it can be difficult to distinguish between causes and symptoms. [3] The report approached these complex issues in a very simple manner, only briefly discussing challenges before offering possible solutions.

The report's causal analysis was helpful in linking causes and their often-unintended effects in a new way. For example, the increasing use of raw materials and chemicals by industry has led to pollution; with a more comprehensive understanding of the causes of pollution, policymakers are better armed in their attempts to decrease its negative effects on human health and the natural environment. [4] Causal analysis also had an educational purpose, in that it helped to persuade the report's readers to adopt more sustainable practices as individuals.

Overlooked

While *Our Common Future* and its main ideas are well studied, its insights into the role of NGOs in promoting sustainable development* is still neglected in the current literature, with a few exceptions. [5] The relationship between NGOs and sustainable development is not so obvious. In fact, radical environmentalists such as the Indian activist Vandana Shiva* have even complained that sustainable development has been harmful for environmental movements, especially in poorer nations, because it risks killing the momentum of more radical struggles to protect the environment. [6] She was aware of the possibility that the introduction of "weak sustainability" * by Western international organizations would overshadow the "strong sustainability" * that environmental NGOs lobbied for. ("Weak sustainability" is a form of sustainability* that sees natural capital*—resources—as potentially replaceable by human capital*—knowledge, technology, and so on; "strong sustainability" is sustainability prioritizing the environment above all.)

Many NGOs contributed to the research for *Our Common Future*, submitting facts they had collected and making suggestions for policy changes. The authors knew it would not be easy for governments to accept policy changes, as the practicalities of sustainable development would require the fundamental reform of the management of resources, technology, institutions, and policy. Change like this is difficult to implement and demands institutional and financial resources, requiring politicians to make "painful choices." [7]

From the beginning, the authors of *Our Common Future* expected to encounter resistance to radical change. They expected that governments would have serious difficulties in addressing environmental crises, because their interests tended to be "too narrow, too concerned with quantities of production or growth." [8] In that sense, sustainable development is a "nongovernmental" topic—that is, governments will not necessarily be the best at making societies more sustainable, and need external pressure and encouragement. In Brundtland's vision, NGOs were pioneers in promoting public awareness and pushing governments to act. [9] However, the report's commissioners have never discussed the weaknesses of NGOs: that they tend to be focused on specific interests and have limited capacity to fund projects without government support.

---

1. William R. Shadish et al., *Foundations of Program Evaluations: Theories of Practice* (London: Sage, 1991), 20–1.
2. World Commission on Environment and Development (WCED), *Our Common Future* (Oxford: Oxford University Press, 1989), ix.
3. David Wasdell, *Studies in Global Dynamics No. 7—Brundtland and Beyond: Towards a Global Process* (London: Urchin, 1987), 7–8.
4. WCED, *Our Common Future*, 2–3.
5. Ben Pile, "Wishing Greenpeace an Unhappy Birthday," *Spiked*, September 12, 2011, accessed February 2, 2016, http://www.spiked-online.com/newsite/article/11068#.VrCGxPkS-Uk.
6. Vandana Shiva, "The Greening of Global Reach," in *Global Ecology: A New Arena of Political Conflict*, ed. Wolfgang Sachs (London: Zed Books, 1993), 149–56.
7. WCED, *Our Common Future*, 9.
8. WCED, *Our Common Future*, 328.
9. WCED, *Our Common Future*, 328.

MODULE 7
ACHIEVEMENT

KEY POINTS
- The importance of *Our Common Future* was recognized all over the world; its recommendations still dominate the political agenda of the United Nations. *
- The report expounded universal values, gaining the sympathy of a wide readership.
- The report has not, however, been directly responsible for any major changes in terms of policy* or in the behavior of individuals.

Assessing the Argument

Gro Brundtland's *Our Common Future*, as its title suggests, was designed to have a general appeal across nations and generations. It aimed to change international and national policies, and individual behavior that damages the environment, as part of "a global agenda for change." [1] In addition to the global character of the problems discussed in the report, the fact that the 21 commissioners came from different nations and were "unanimous in our conviction that the security, well-being, and very survival of the planet depend on such changes" [2] was seen as proof that the report was relevant throughout the world.

Our Common Future set out to address the major challenges faced by the global community in the 1980s. At the time of publication in 1987, the authors hoped that these challenges could be resolved by the year 2000. [3] However, the actions of policymakers and the general public in response to the report were not sufficient to achieve the target in time. This can be demonstrated by the fact that all the problems mentioned in the report, such as pollution, population growth, economic crises, poverty, and climate change, * are still facing us today and have even worsened over time. For this reason, the problems discussed in the report are very familiar to today's readers.

Our Common Future noted the global nature of the issues under discussion, and how they applied to both developed and developing countries—that is, both wealthy and impoverished countries. 4 However, the specifically limited capabilities of developing nations in areas such as knowledge and economic resources were also discussed and an argument was made in favor of increasing the aid given to impoverished countries by developed countries. 5 Some scholars and activists, among them the Swiss economist Matthias Finger, * criticized the distinction made in the report between wealthy developed countries and impoverished developing countries. They argued that making distinctions like this weakened and divided the environmental ("Green") movement. 6

> "I remember I gave a kind of colder warning about the results so far, saying that, yes, there is progress in many fields, there is some progress in some fields, but also in several areas no progress at all."
>
> ——Gro Brundtland, *Interview with Peter Ocskay,* Baltic University, Uppsala, Sweden, 1997

Achievement in Context

Just three years after the publication of *Our Common Future*, the Canadian minister of the environment Lucien Bouchard* stated that the concept of sustainable developmen* as introduced by the report "has changed forever the way we think about the environment." [7]

In the decades since publication, the concept of sustainable development has been widely accepted, and has become an important framework for policymakers. Not only did the idea of sustainable development produce the new discipline of environmental economics, *[8] it has also shaped the policymaking processes at all possible levels and become a common political term.

By 2009, the sustainable development agenda was incorporated into national development* strategies in 106 countries. The business sector had accepted a new agenda, corporate and social responsibility* (CSR), although some businesses who took this up were criticized for "greenwashing"* —a marketing spin by businesses and organizations wanting to promote themselves as environmentally friendly. [9] Criticism of environmentally damaging business practices has led to some businesses collaborating with the environmental and social NGOs* forming joint initiatives to address sustainable development. National governments remain slow in reaching binding (compulsory) agreements about sharing responsibility for global matters, such as pollution, climate change, and food insecurity. *

The report's aim of promoting multilateral* collaboration, then, has partly been met. From this perspective, the report was a game changer for national and international policy, going far beyond the expectations of its authors.

Limitations

In the 1980s, the World Commission on Environment and Development* headed by Brundtland was hoping to achieve sustainable development by the year 2000. [10] However, despite all the actions suggested, the world has yet to solve the problems addressed in *Our Common Future* because "tensions, controversies and gridlocks between development and environment still exist." [11]

Despite worldwide recognition of the concept of sustainability, * little progress has been made in achieving it. The measures suggested by *Our Common Future* were voluntary. [12] Institutional follow-ups to the report, including the UN's Earth Summit* in 1992 in Rio de Janeiro and the Rio+20 Conference* in 2012, faced serious problems in "designing the move from theory to practice." [13] The main challenge is that "a huge constituency around the world cares deeply and talks about sustainable development, but has not taken serious on-the-ground action." [14] Similar feelings have been shared by policy analysts about the 1997 Kyoto Protocol* (an international treaty that aimed to slow climate change by reducing the emission of certain gases understood to cause it) and the Paris Conference* on climate change held in 2015. [15]

At a theoretical level, dissecting the concept of sustainable development into three parts (economic, social, and environmental development*) has led to another challenge: finding a balance between the environment and the economy. This challenge provokes fundamental debates on what should be prioritized: economic growth, as in the weak sustainability* model, or environmental conservation, as in the strong sustainability* model. [16]

---

1. World Commission on Environment and Development (WCED), *Our Common Future* (Oxford: Oxford University Press, 1989), ix.
2. WCED, *Our Common Future*, 343.
3. WCED, *Our Common Future*, ix.
4. WCED, *Our Common Future*, 52.
5. WCED, *Our Common Future*, 60.
6. Matthias Finger, "Politics of the UNCED Process," in *Global Ecology: A New Arena of Political Conflict*, ed. Wolfgang Sachs (London: Zed Books, 1993), 36.
7. Linda Starke, *Signs of Hope: Working towards Our Common Future* (Oxford: Oxford University Press, 1990), 21.
8. Herman Daly, *Ecological Economics and Sustainable Development: Selected Essays* (Cheltenham: Edward Elgar, 2007), 251.
9. John Drexhage and Deborah Murphy, *Sustainable Development: From Brundtland to Rio 2012* (New York: United Nations, 2010), 15.
10. WCED, *Our Common Future*, ix.
11. Volker Hauff, "Brundtland Report: A 20 Years Update" keynote speech, European Sustainability, Berlin, June 3, 2007, accessed February 1, 2016, http://www.nachhaltigkeitsrat.de/uploads/media/ESB07_Keynote_speech_Hauff_07-06-04_01.pdf.
12. Jennifer Elliott, *An Introduction to Sustainable Development: The Developing World* (London: Routledge, 2000), 8.
13. Elliott, *Introduction to Sustainable Development*, 8.
14. Drexhage and Murphy, *Sustainable Development*, 2.
15. Thomas Sterner, "The Paris Climate Change Conference Needs to Be More Ambitious," *Economist*, November 18, 2015, accessed February 2, 2016, http://www.economist.com/blogs/freeexchange/2015/11/cutting-carbon-emissions.
16. Drexhage and Murphy, *Sustainable Development*, 10.

MODULE 8
PLACE IN THE AUTHOR'S WORK

KEY POINTS

- Gro Brundtland made an important contribution to the discussion of contemporary global challenges, including environmental degradation, climate change, * and human health.
- *Our Common Future* provided a revolutionary approach to human development* in the 1980s and remains the best-known work on sustainability* despite recent criticisms; Gro Brundtland has continued to promote sustainable development. *
- *Our Common Future* is a milestone work in the careers of all the Brundtland commissioners, * including Brundtland herself.

Positioning

By the time Gro Brundtland was invited to chair the World Commission on Environment and Development, * she had already won a reputation as an experienced politician. Not only had she served two terms as the first female prime minister of Norway (1986–9 and 1990–6), she was also the only prime minister to have served as minister of the environment (1974–9).

This unique background allowed Brundtland to develop new ideas for the management of finite resources and the environment, and human development. She was already well known before the publication of *Our Common Future*, but the commission's report bolstered her status as a celebrity politician. She was subsequently invited to write short texts capturing major moments in the commission's history, including the foreword to a book about the writing of the report, and she has given numerous public speeches. However, since then she has not produced a work equal to the significance of *Our Common Future*. [1]

In 1997–8, Brundtland wrote two autobiographical works in Norwegian, the titles of which translate as *My Life* [2] and *Dramatic Years 1986–1996*. [3] In 2002 she wrote an autobiography in English entitled *Madam Prime Minister: A Life in Power and Politics*. [4] The book describes major moments of her life and her encounters with other prominent leaders, such as the French president Francois Mitterand, * the French statesman Jacques Delors, * the first president of the Russian Federation Boris Yeltsin, * the US politician Hillary Clinton, * and the South African freedom fighter and president Nelson Mandela. * Her husband, Arne Olav Brundtland, also a Norwegian politician, wrote two memoirs capturing their lives during Gro Brundtland's public service. Their titles translate into English as *Married to Gro* and *Still Married to Gro*. [5] During Brundtland's tenure as director-general of the United Nations* health body, the World Health Organization* (WHO), she wrote about the influence of environmental factors and climate change on human health. Recently she has publicly discussed the role of democracy and human rights in achieving sustainable development.

> *"In the end, I know I made the right choice of taking the challenge the UN gave me. It has had a great impact on this country [Norway] and globally."*
>
> —— Gro Brundtland, *Interview with the Uongozi Institute*

Integration

Brundtland has remained in favor of *Our Common Future*'s main argument: that there is an important link between the management of environmental resources and economic growth. While she believes that the Northern hemisphere's development issues have almost been fixed, she has raised concerns about the mass poverty of developing countries, which has only seemed to worsen since the publication of the report. This point is also often raised by prominent critics of the report, such as the Indian environmental activist Vandana Shiva*[6] and the Swiss economist Matthias Finger.*[7] *Our Common Future* discussed the universal responsibility of humankind to nature and the necessity of cooperation between the Global North* and the Global South*; in her later writings Brundtland went even further in arguing that Northern consumers are partly responsible for the impoverishment and environmental degradation of the Global South, which could cause future conflicts. [8]

Brundtland used a similar approach to that of *Our Common Future* in her later works during her service in WHO. In 1998, she published "Macroeconomics and Health," with the economist Jeffrey D. Sachs, * which argued that health and longevity were both fundamental goals of development and the means to achieving development goals, and referenced the concept of sustainable development. [9] The WHO report also linked health to poverty reduction and long-term economic growth.

It can be argued that Brundtland has maintained the views she expressed in the report regarding economic growth as a means of achieving international development goals. The negative environmental impact of human development as a consequence of human economic activity was not discussed in the report—and nor has Brundtland mentioned it since.

Significance

Our Common Future is Brundtland's best-known work by far. When she began the process of chairing the commission that produced the report, she had already earned a good reputation in UN circles and this enhanced the commission's political credibility. The report made Brundtland internationally famous, as she was responsible for formulating arguably one of the most important concepts of modern political economy—sustainable development.

Brundtland, today a member of the Elders* (a think tank* made up of retired senior UN officials and former political leaders), has remained loyal to the ideas and ideals presented in *Our Common Future*. She has echoed them in her later, lesser-known works and has carried on promoting the concept of sustainable development in the years since the publication of the report, notably during her tenure as the director-general of WHO.

While the new ideas presented in *Our Common Future* were well received by most governments and by the general public in the 1980s, [10] more recent interpretations of the report and its core concept of sustainable development have been more critical. Environmental activists and political economists have referred to the report as an example of neoliberal* thinking, arguing that it relies on the false possibility of infinite economic growth based on efficient use of resources through technological advancement and international cooperation. [11] ("Neoliberalism" is an approach to economics that calls for the market to operate without hindrance from regulation or government intervention, and without regard to any social consequences.) Although Brundtland acknowledged and agreed with this critique in a public speech in Vienna in 2013, she has not published anything on the matter.

---

1. See Linda Starke, *Signs of Hope: Working towards Our Common Future* (Oxford: Oxford University Press, 1990).
2. Gro Harlem Brundtland, *Mitt Liv: 1939–1986* (Oslo: Gylenhal, 1997).
3. Gro Harlem Brundtland, *Dramatiske ar 1986–1996* (Oslo: Gyldedal, 1998).
4. Gro Harlem Brundtland, *Madam Prime Minister: A Life in Power and Politics* (New York: Farrar, Straus and Giroux, 2005), 19.
5. See Arne Olav Brundtland, *Gift med Gro* (Oslo: Oslo University, 1996) and Arne Olav Brundtland, *Fortsatt Gift med Gro* (Oslo: Oslo University, 2003).
6. Vandana Shiva, "The Greening of Global Reach," in *Global Ecology: A New Arena of Political Conflict*, ed. Wolfgang Sachs (London: Zed Books, 1993), 149–56.
7. Matthias Finger, "Politics of the UNCED Process," in *Global Ecology: A New Arena of Political Conflict*, ed. Wolfgang Sachs (London: Zed Books, 1993), 36.
8. Gro Harlem Brundtland, "Global Change and Our Common Future: The Benjamin Franklin Lecture," in *Global Change and Our Common Future*, ed. R. S. DeFries and T. F. Malone (Washington DC: National Academy Press, 1989), 15.
9. Gro Harlem Brundtland and Jeffrey D. Sachs, "Macroeconomics and Health: Investing in Health for Economic Development," (Geneva: World Health Organization, 2001), 3.
10. Starke, *Signs of Hope*, 2.
11. Gearoid O Tuathail et al., eds, *The Geopolitics Reader* (London: Routledge, 1998).

SECTION 3
IMPACT

MODULE 9
THE FIRST RESPONSES

KEY POINTS
* While *Our Common Future* and its main concept, sustainable development, * received a generally positive response, the main critiques came from environmental activists and economists who criticized the report for adopting an approach known as "weak sustainability."*
* None of the report's authors have entered into a public debate with their critics.
* Gro Brundtland's later works partly respond to criticisms by introducing a fourth element, culture, into the three-part model of sustainability given in the report.

Criticism

Although the recommendations for raising environmental awareness made in Gro Brundtland's *Our Common Future* were generally welcomed by scientists and policymakers, * some aspects of the new concept of sustainable development were criticized. One of the first critiques came from environmental activists and economists, who argued that the concept of sustainable development as described in the report was very broad and did not specify what exactly had to be sustained. In the report, sustainability was used both to refer to the responsible management of natural resources and as an economic term. For example, the authors used the concept of sustainability in relation to foreign debt, which was described as "unsustainable" when it results in the flight of local capital from developing countries (that is, when money from a less-wealthy nation is invested elsewhere). [1]

Generally, in the report, sustainable development was equated with sustainable growth, promising economic prosperity under a new order of responsible management of natural resources. 2

The Indian social scientist Shiv Visvanathan* has questioned how realistic it is to combine the concepts of environmental sustainability and development. * He argued that the type of development imposed by international organizations such as the United Nations* was "a genocidal act of control" forced upon developing countries and did not adequately address real ecological issues. 3

The way that *Our Common Future* linked the state of the environment and economic development was innovative at the time. But ecological economists* termed this approach "weak sustainability," because it saw innovation and knowledge as a possible substitute for natural capital* (natural resources useful to human life and economics). 4 For them, the approach gave "a sense of comfort to the effect that the environment can be dispensed with." Instead they demanded "strong sustainability" —an approach that gave more emphasis to the protection of natural resources. 5

> *"In my final response I just exclaimed: 'Yes, all is indeed linked to everything else.' At the time, 20 years ago, this final statement from me as prime minister was immediately criticized by the opposition, even ridiculed by some, for being evasive and unclear. Interestingly, today it is often quoted with great respect and even admiration."*
>
> —— Gro Brundtland, *Climate Change and Our Common Future*, speech at the Global Climate and Energy Project (GCEP) Symposium

Responses

Brundtland has maintained the position she expressed in the report about the need for sustainable development. Over the years she has deepened her understanding of the concept of sustainability by adding issues of democracy and human rights. [6] The report has been criticized as an example of how greenwashing* (superficial support for ecological issues without effecting real change, often in the name of material gain) can damage the global environmental movement, as argued by the Indian environmental activist Vandana Shiva. [7] However, neither Brundtland, nor any of the other commissioners who helped to produce the report, have engaged in a public discussion with their critics on this matter.

Our Common Future discussed the universal responsibility of humankind to protect nature and the necessity of cooperation between the Global North* and the Global South. * In her later writings, Brundtland went even further in arguing that Northern consumers played a role in the impoverishment of the Global South, which could serve as a source of future conflicts. [8] This could be seen as a response to her critics.

It is not known how much Brundtland agrees with the ecological economists who critiqued the concept of sustainable development, disagreeing with the report's argument that natural capital can be replaced by the knowledge, technology, and innovation known as "human capital." No one, it seems, has thought to ask her this question directly. She is known to be a cautious politician and is careful about the public platforms and writing opportunities she chooses. [9]

Conflict and Consensus

The concept of sustainability presented in *Our Common Future* was criticized by economists and environmental activists for promoting economic growth over environmental protection. It was also criticized for using a neoliberal* model of development in which the market must be allowed to operate without hindrance, and according to which the Global North dictates to the Global South, ignoring or destroying local environmental movements. [10]

However, the authors of the report believed that their main goal was to raise awareness about environmental and social crises, and not to recommend an exact set of actions. Like Brundtland, many of the authors were politicians themselves, and their status kept them from suggesting radical reforms to the world order. In fact, they considered the idea of sustainability to be radical enough and were aware of the political risks involved in pushing for further reforms.

Our Common Future and Brundtland's speeches indicate that the concept of sustainability was centered on people and their needs and she did not see anything wrong with that. Humanity and nature were seen as an interlinked system: "People influence the trends that [decide the destiny of] the planet. The planet affects people." [11] Brundtland believed that focusing on human needs was beneficial for both people and the planet.

The main compromise that Brundtland has made in response to critiques of the report was introducing a fourth element into the three-pillar system of sustainability: culture. [12] She added the concept of cultural relativism—the idea that different cultures have different values and practices that cannot be compared directly to each other— to the concept of sustainability. This allowed her to explain why different countries or regions introduced sustainability at different speeds and in different ways. This new addition avoided the one-size-fits-all approach for which the report had previously been criticized.

---

1. World Commission on Environment and Development (WCED), *Our Common Future* (Oxford: Oxford University Press, 1989), 73.
2. WCED, *Our Common Future*, 68.
3. Shiv Visvanathan, "Mrs Brundtland's Disenchanted Cosmos," *Alternatives* 16 (1991): 378–81.
4. David Pearce and Giles Atkinson, "The Concept of Sustainable Development: An Evaluation of its Usefulness Ten years after Brundtland," *Swiss Journal of Economics and Statistics* 134, no. 3 (1998): 4–5.
5. P. A. Victor et al., "How Strong is Weak Sustainability?" *Economie Appliquée* 48, no. 2 (1995): 75–94.
6. Gro Harlem Brundtland, *Opening Speech to the 2015 Nobel Peace Prize Forum*, Oslo, March 10, 2015, accessed February 1, 2016, https://www.youtube.com/watch?v=LBRmSjsnVGs.
7. Vandana Shiva, "The Greening of Global Reach," in *Global Ecology: A New Arena of Political Conflict*, ed. Wolfgang Sachs (London: Zed Books, 1993), 149–56.
8. Gro Harlem Brundtland, "Global Change and Our Common Future: The Benjamin Franklin Lecture," in *Global Change and Our Common Future*, ed. R. S. DeFries and T. F. Malone (Washington DC: National Academy Press, 1989), 15.
9. David Wilsford, *Political Leaders of Contemporary Western Europe: A Biographical Dictionary* (Westport, CT: Greenwood Press, 1995), 55.
10. Visvanathan, "Mrs Brundtland's Disenchanted Cosmos," 378–9.
11. Gro Harlem Brundtland, "Healthy People, Healthy Planet," the Annual Lecture in the Business and the Environment Program, London, March 15, 2001, accessed February 1, 2016, http://www.cisl.cam.ac.uk/publications/archive-publications/brundtland-paper.
12. Gro Harlem Brundtland, "ARA Lecture," Vienna Technical University, November 18, 2013, accessed February 1, 2016, https://www.youtube.com/watch?v=X7Z2o8tZZoE.

MODULE 10
THE EVOLVING DEBATE

KEY POINTS

- Environmental activists and economists have criticized *Our Common Future* and the concept of sustainability* for focusing too much on economic growth.
- United Nations (UN)* officials and the Brundtland commissioners* have recognized that progress toward achieving sustainability has been slow.
- A new generation of activists and progressive economists from developing nations have offered alternative frameworks to those in the report.

Uses and Problems

Our Common Future was researched and written collaboratively under the leadership of Gro Brundtland. Its central concept of sustainable development* has been at the core of contemporary debates on human development* ever since its publication; it continues to be the main framework for analysis of many contemporary global matters. The report remains useful in reminding current leaders about the complex nature of the link between the natural environment and international development.

In 2012, the UN held a conference on sustainable development in the Brazilian city of Rio de Janeiro, Rio+20, * at which national governments were offered a non-binding agreement known as *The Future We Want*. While some governments voluntarily committed to achieving sustainable development, as set out in the agreement, the document's lack of compulsory commitments led to widespread disappointment with the outcomes of the conference; media and civil society* saw it as "a caricature of diplomacy." [1]

The gravity of the current problems in international development and the lack of progress in tackling them have also shifted perspectives regarding *Our Common Future*. The general attitude toward the report and its optimistic account of the world's capacity to successfully address the economic, environmental, and social crises has changed from initial praise and optimism to more critical and sometimes pessimistic opinions. Some, such as the Slovenian social theorist Drago Kos, * have expressed strong doubts about the world's ability to ever achieve sustainability. [2]

> *"While sustainable development is often perceived as an environmental issue, it has been subject to competing agendas."*
> —— John Drexhage and Deborah Murphy, *Sustainable Development: From Brundtland to Rio 2012*

Schools of Thought

Initially, all contributors to *Our Common Future* agreed about the need to manage environmental resources sustainably. Later, the concept of sustainable development was debated by proponents of strong sustainability,* with its emphasis on environmental protection, and others who argued for weak sustainability,* which acknowledged the importance of economic development. *[3] Supporters of the weak sustainability approach include Brundtland commissioners, such as the German politician Volker Hauff* and the Canadian businessman and diplomat Maurice Strong, * who argued that lack of economic growth has caused environmental deterioration. This approach is naturally supported by governments and corporations that tend to prioritize economic growth above all.

In contrast, environmental or ecological economists such as Herman Daly, * a supporter of strong sustainability, argue that the protection of natural resources should be given priority and that doing otherwise will lead to further economic deterioration. The school of ecological economics has embraced the concepts of strong sustainability and weak sustainability and integrated them into its discussions regarding the place of natural capital* in the management of resources. [4]

In Current Scholarship

The arguments made by environmental economists and environmental activists about the benefits of strong sustainability over weak sustainability are still being debated.

The most recent version of the strong sustainability argument is an approach called "degrowth," which means rejecting ideas of neoliberal* economic growth. The concept was invented by grassroots environmental movements, particularly in South America, and was first presented in an academic framework by the Columbian American anthropologist Arturo Escobar. * The concept reformulates the meaning of economic growth. Under the degrowth model, the economy can be allowed to grow only in the sectors that benefit people and their livelihoods, such as health and education. [5]

This model promotes the strong sustainability approach, as it argues for the protection of nature. An example of how this model is realized in practice is the "Universal Declaration of the Rights of Mother Earth," [6] presented in Cochabamba, Bolivia, in April 2010. This was an alternative to the "Earth Charter" that failed to be adopted at the Earth Summit* in 1992. [7] Bolivia has pioneered the way in passing this document, which put the concept of the earth's rights on the UN agenda.

Ecological economists* also continue to criticize the model of economic growth based on production and consumption, especially when it overtakes the rate at which natural resources can replenish themselves. For example, the Harvard economist Michael Porter*[8] envisions a more efficient use of resources based on technological innovation, a win-win situation that would make the overexploitation of natural resources unnecessary. This notion is in line with *Our Common Future*, which considered technology one of the solutions to global challenges. Based on Porter's work, a new generation of economists[9] has offered a comprehensive framework to build a sustainable economy. This framework includes new business models based on ethical values and less focus on economic profit, promotes community-based ownership structures such as cooperatives (businesses or services owned by the people who work at them), and aims to achieve better measures of socioeconomic progress (that is, measures of when things are getting better socially and economically).

1. Jim Leape, "It's Happening, but Not in Rio," *New York Times*, June 24, 2012, accessed February 1, 2016, http://www.nytimes.com/2012/06/25/opinion/action-is-happening-but-not-in-rio.html?_r=0.
2. Drago Kos, "Sustainable Development: Implementing Utopia?" *SOCIOLOGIJA* 54, no. 1 (2012): 7–20.
3. Jerome Pelenc, "Weak Sustainability Versus Strong Sustainability," Brief for *GSDR* (Louvain: UCL, 2015), 1–4.
4. John M. Gowdy and Marsha Walton, "Sustainability Concepts in Ecological Economics," *Economics Interactions with Other Disciplines*, vol. 2. *Encyclopedia of Life Support Systems* (Paris: UNESCO/Eolss, 2008), 111–20.
5. Arturo Escobar, "Alternatives to development," interview with Rob Hopkins, Venice, September 28, 2012, accessed February 1, 2016, http://transitionculture.org/2012/09/28/alternatives-to-development-an-interview-with-arturo-escobar/.
6. Global Alliance for the Rights of Nature, "Universal Declaration of the Rights of Mother Earth," Cochabamba, Bolivia, April 22, 2010, accessed December 14, 2015, http://therightsofnature.org/universal-declaration/.
7. Maurice Strong, "The 1992 Earth Summit: An Inside View," interview with Philip Shabecoff, Quebec, 1999, accessed February 1, 2016, http://www.mauricestrong.net/index.php/earth-summit-strong.
8. Michael E. Porter and Claas van der Linde, "Green and Competitive: Ending the Stalemate," *Harvard Business Review* 73, no. 5 (1995).
9. D. W. O'Neill et al., "Enough Is Enough: Ideas for a Sustainable Economy in a World of Finite Resources," *Report of the Steady State Economy Conference* (Leeds: CASSE and Economic Justice for All, 2010), 9.

MODULE 11
IMPACT AND INFLUENCE TODAY

KEY POINTS
* *Our Common Future* remains significant today in both academic thinking and public policy. *
* The concept of sustainability* presented in *Our Common Future* has been criticized for being difficult to achieve in practice.
* Gro Brundtland still works in international politics as a senior policy adviser and remains committed to the concept of sustainable development. *

Position

Gro Brundtland's *Our Common Future* is the definitive source for the concept of sustainable development, and remains at the heart of current debates regarding the future path of development. *¹

One major obstacle in the way of achieving sustainable development is how fragmentary the required actions are, involving not just changes in the behavior of national governments and how they work together, but in the behavior of individuals. Another is the ambiguity of the concept of sustainable development, and the complexity of the issues it covers. ² Dealing with the future of human development, the topic has very high stakes; consequently, the debate about sustainable development has gone well beyond the narrow confines of those with an interest in the field of ecological economics. It is an issue that concerns us all.

The United Nations (UN), * the institution that originally commissioned the *Brundtland Report*, has never questioned the viability of sustainable development. Instead, it has tried to incorporate critiques and adapt the concept to new challenges. UN agencies took on the role of putting sustainable development into practice. The first institutional follow-up to the report was at the Earth Summit* in 1992 when UN officials, including Maurice Strong, * the secretary-general of the conference and a Brundtland commissioner, * tried to formulate the necessary policy changes in a concrete set of actions known as "Agenda 21." * Another attempt at international agreement on sustainable development took place 20 years later in Rio de Janeiro at the Rio+20 Conference. * This resulted in an agreement that the Millennium Development Goals, * a list of aims for human development around the world, would include sustainability targets and be referred to as the Sustainable Development Goals (SDGs). *

> "Sustainable development is a visionary development paradigm, and over the past 20 years governments, businesses, and civil society have accepted sustainable development as a guiding principle ..."
> —— John Drexhage and Deborah Murphy, *Sustainable Development: From Brundtland to Rio 2012*

Interaction

There have been several UN conferences to promote the implementation of sustainable development as recommended in *Our Common Future*. Of these, the Earth Summit of 1992 and the Rio+20 Conference of 2012 were particularly important. The summits were criticized for putting too much emphasis on environmental issues, and paying too little attention to development aid from wealthy to impoverished countries and to cooperation between countries. [3]

Gro Brundtland and Maurice Strong, both high-profile members of the Brundtland Commission, have remained engaged in the debate on sustainability for more than 20 years. Despite heavy criticism of the slow progress of sustainability, Brundtland and Strong stayed loyal to the concept, believing that the barriers to sustainability did not reflect fundamental flaws in the concept but were attributable to other factors.

Brundtland recognized that the challenge of sustainability was bigger than the system that was supposed to implement it. According to her recent speeches, international institutions could make better progress towards achieving it. Strong has also referred to institutional weaknesses of the UN that made it difficult for the organization to provide strong global leadership on sustainability. But he was highly appreciative of the attempts made by Boutros Boutros-Ghali, * UN secretary-general between 1992 and 1996, to reform the UN, "bring[ing] together several of the elements within the secretariat into a new economic and social affairs department" called the Economic and Social Council (ECOSOC), * a body responsible for economic and social activities. [4] ECOSOC was given a central role in following up the decisions made at the Rio Conference.

Strong served as a senior adviser at the Rio+20 Conference, and used this opportunity to suggest institutional innovations to the UN system; among these were restructuring the United

Nations Environment Program (UNEP), * the UN body responsible for coordinating the organization's environmental activities and policies, and offering assistance to developing nations seeking to implement practices and policies designed to protect the environment. [5] Despite criticisms, the Rio+20 Conference succeeded in reformulating the Millennium Development Goals as Sustainable Development Goals (SDGs), ensuring that sustainability targets were included in official international development aims. This was an important achievement.

The Continuing Debate

The debate on how to make sustainability work in practice continues. While some, among them the Slovenian social scientist Drago Kos, *[6] argue that the concept is utopian (that is, it belongs in an impossibly perfect world), most scientists and politicians assume that it is possible to make sustainable development a reality. Many others, including the South African religious leader and activist Desmond Tutu, * insist that extra effort has to be made if humanity is to survive in the near future.

UN officials and Brundtland commissioners are aware of the slow progress on implementing sustainability. In 2010, the UN High Level Panel on Global Sustainability, a committee of experts convened to discuss the subject, commissioned scientists from the International Institute for Sustainable Development* (an independent, nonprofit organization founded to promote human development by sustainable means) to study the causes of this slow progress and possible solutions. Certain development specialists argue for deep institutional changes in global economics, particularly in the area of global business, in order to make sustainable development happen. [7]

Brundtland, herself a member of the UN High Level Panel on Global Sustainability, has also addressed this question. In a public lecture at Vienna University of Technology she gave her answer to the following question: "Why are we not succeeding in changing our ways and building a sustainable future 20 years after Rio?" [8] She thinks that there are political, social, and technological reasons for the slow progress, and the inefficiency of global systems of government is, for her, the main reason that sustainable development has not been sufficiently widely adopted. [9]

---

1. Volker Hauff, "Brundtland Report: A 20 Years Update" keynote speech, European Sustainability, Berlin, June 3, 2007, accessed February 1, 2016, http://www.nachhaltigkeitsrat.de/uploads/media/ESB07_Keynote_speech_Hauff_07-06-04_01.pdf.
2. John Drexhage and Deborah Murphy, *Sustainable Development: From Brundtland to Rio 2012* (New York: United Nations, 2010), 16.
3. Maurice Strong, *Where on Earth Are We Going?* (Toronto: Vintage Canada, 2001), 78.
4. Maurice Strong, "The 1992 Earth Summit: An Inside View," interview with Philip Shabecoff, Quebec, 1999, accessed February 1, 2016, http://www.mauricestrong.net/index.php/earth-summit-strong.
5. Maurice Strong, "Statement by Maurice F. Strong delivered to the Special United Nations General Assembly Event on Rio+20," New York, October 25, 2011, accessed February 1, 2016, http://www.unep.org/environmentalgovernance/PerspectivesonRIO20/MauriceFStrong/tabid/55711/Default.aspx.
6. Drago Kos, "Sustainable Development: Implementing Utopia?" *SOCIOLOGIJA* 54, no. 1 (2012): 20.
7. Drexhage and Murphy, *Sustainable Development*, 19–20.
8. Gro Harlem Brundtland, "ARA Lecture," Vienna Technical University, November 18, 2013, accessed February 1, 2016, https://www.youtube.com/watch?v=X7Z2o8tZZoE.
9. Brundtland, "ARA Lecture."

MODULE 12
WHERE NEXT?

KEY POINTS

- It is very likely that the definition of sustainable development* established in *Our Common Future* will remain in use for the foreseeable future.
- It is also likely that the understanding of sustainability* founded on the three pillars of environment, human development, * and economic growth will be expanded into a multidimensional model.
- *Our Common Future* is a groundbreaking, highly praised text that has provided the most widely used definition of "sustainable development."

Potential

Humanity is likely to carry on facing the challenges identified by *Our Common Future*, such as climate change, * overpopulation, and decreasing natural resources. Sustainability will remain at the forefront of the international political agenda, and it is likely that policymakers* will continue to draw upon the classic definition of sustainability as formulated by Brundtland.

The issues discussed in the report continue into the twenty-first century, and many have worsened. It is expected that by 2050 the world's population will reach 9.1 billion, and the main increase will occur in developing countries. [1] In order to feed this population and avoid food insecurity, * food production will have to increase by 70 percent. [2] Balancing this increased demand in food with rapidly decreasing natural resources, such as fertile land and water, will be difficult and will require a serious change in resource management and the political will to bring this about. The solutions offered in *Our Common Future*, such as technological development and international cooperation, will definitely have to be applied.

Failure to find the right solution could lead to a serious global crisis that could threaten the very survival of the human species. As the South African religious leader and activist Desmond Tutu* put it, "We have only one world, and if we destroy it, we are done." [3]

> "The fact is a compelling reality, and should not easily be dismissed. Since the answers to fundamental and serious concerns are not at hand, there is no alternative but to keep on trying to find them."
> ——World Commission on Environment and Development,*
> *Our Common Future*

Future Directions

Our Common Future proposed a revolutionary approach to the management of natural resources by connecting the environment with human development and adding the third element of economic growth. These are the three pillars that make up the concept of sustainable development that has served well in explaining the complexities of human development, although the concrete meaning of each of these pillars is still disputed.

To address this issue, a number of political economists and social scientists, including philosophers, have started to offer multidimensional models of sustainable development. For example, the Colombian American anthropologist Arturo Escobar* has offered an alternative development concept that "take[s] into account place-based models of nature, culture, and politics." [4] Culture is the fourth element to be added to the original three-part model of sustainability.

Another dimension that can be added to the model is global governance—the means of political cooperation between governments. It is clear, even from the example of *Our Common Future* itself, that innovative ideas about the future of mankind will remain just ideas unless there are concrete mechanisms to make them a reality. The report rightly predicted that national governments would be reluctant to shift government policy toward sustainability, and emphasized the importance of lobbying from civil society*—the general public—to make it happen. However, there are gaps in the current understanding of how the role of civil society can be optimized in this process, so this question will continue to inspire theories and debates.

Summary

Our Common Future is a report prepared by the United Nations* World Commission on the Environment and Development* in 1987. It is also known as the *Brundtland Report*, named after the chairperson of the Commission, the former Norwegian prime minister Gro Harlem Brundtland. The report is considered the definitive source text for the concept of sustainable development. Although some earlier works mentioned a similar concept, it was *Our Common Future* that popularized the idea of a link between human development and natural resources, arguing for responsible management of natural resources in order to maintain them for future generations. This has become accepted today as the definition of sustainable development. It is frequently used in both academic work and political contexts.

Our Common Future introduced the conceptual cornerstones of contemporary development policies: the three pillars of economic growth, social development, and natural and ecological conservation. The report received much positive attention immediately after publication and was hailed by Oxford University Press as "the most important document of the decade on the future of the world." Indeed, it is possible to expand this point by claiming the report to be one of the most influential texts of the twentieth century. As the challenges and the possible solutions discussed within it have remained relevant, the report continues to influence political thinking in the twenty-first century.

Most international organizations, including the UN, continue to include sustainable development in their agenda. The recent formulation and adoption of the UN's Sustainable Development Goals, * updating the Millennium Development Goals* adopted in New York in 2000, is another example of how sustainability remains relevant in the twenty-first century. There is no doubt that *Our Common Future* will remain a highly influential text for decades to come.

---

1. Food and Agriculture Organization (FAO), "How to Feed the World in 2050" (Rome: Food and Agriculture Organization, 2009), 2.
2. FAO, "How to Feed the World," 2.
3. Desmond Tutu, "Is Sustainable Development a Luxury We Can't Afford?," interview with the Elders, Cape Town, May 12, 2012, accessed February 1, 2016, https://www.youtube.com/watch?v=ArHey8SVJQE.
4. Arturo Escobar, "Culture Sits in Places: Reflections on Globalism and Subaltern Strategies of Localization," *Political Geography* 20 (2001): 139–74.

GLOSSARY OF TERMS

1. **Agenda 21:** an action plan to implement sustainable development principles at both international and national levels, developed by the UN Conference on the Environment and Development in Rio de Janeiro in 1992. The 21 in its title refers to the twenty-first century. It was a voluntary agreement. While European countries strongly supported the plan, the United States opposed it (which explains why it was never fully implemented).

2. **Bhopal disaster:** a gas leak at a pesticide factory in the Indian city of Bhopal in December 1984, it is considered the worst-ever industrial disaster. At least 3, 787 deaths were confirmed in the Madhya Pradesh region where the accident happened.

3. **Brandt Report:** a report drafted by the Independent Commission on International Development, chaired by the West German political leader Willy Brandt, dealing with global economic development.

4. **Brundtland Commission:** formally titled the World Commission on Environment and Development, a body convened by the United Nations with the aim of fostering cooperation between nations to pursue sustainable development.

5. **Causal analysis:** a method of studying the causes of studied effects.

6. **Chernobyl accident:** an industrial accident in April 1986 at a nuclear power-plant in what is today Ukraine, then the Soviet Union. It was the worst nuclear power-plant accident the world has known, affecting thousands of people.

7. **Civil society:** the part of society consisting of everything except the state and business sectors.

8. **Climate change:** patterns of change in global weather conditions resulting from natural and man-made causes, linked to global warming.

9. **Club of Rome:** an international think tank founded in 1968 at the Academia dei Lincei in Rome. Its members include international bureaucrats, political leaders, and prominent businessmen. The Club actively participates in debates on international development.

10. **Cold War (1947–91):** a period of political tension between the United States and the Soviet Union, and their respective allies. While the two countries never

engaged in direct military conflict, they engaged in covert and proxy wars and espionage against one another.

11. **Corporate and Social Responsibility (CSR):** a system of self-regulation by corporations, based on ethical principles. It promotes a "triple bottom line" of people, planet, and profit.

12. **Cuban missile crisis:** a confrontation between the United States and the Soviet Union in October 1962, provoked by the installation of nuclear weapons in Cuba. This was the moment when the world came closest to a nuclear war.

13. **Development:** a network of political and social processes aimed at increasing people's life choices; mostly used as a term by governments and other organizations that aid, manage, and regulate societies.

14. **Earth Summit:** a United Nations conference held in Rio de Janeiro, in June 1992. It promoted the Rio Declaration on the Environment and Development and Agenda 21.

15. **Economic and Social Council (ECOSOC):** one of the principal bodies of the United Nations system, responsible for economic and social activities.

16. **Ecosystem:** a biological system comprising all the organisms that exist in a specific physical environment.

17. **Elders:** an independent group of retired world leaders who work together for peace and human rights, brought together by the South African statesman Nelson Mandela in 2007.

18. **Environmental economics:** the study of economic matters as they are affected by issues of sustainability, and of the economic consequences of policies designed to protect the environment.

19. **Food insecurity:** a lack of food supply leading to malnutrition, hunger, and starvation. It is a condition often present in developing countries.

20. **General Assembly:** one of the main organizations of the United Nations and the only one in which all member nations have equal representation.

21. **Global North:** a global socioeconomic and political category comprising North

America and Western Europe.

22. **Global South:** a global socioeconomic and political category comprising Africa, South America, Asia, and the Middle East.

23. **Greenwashing:** a form of marketing in which an organization pretends to be environmentally friendly while not making any real changes.

24. **Human capital:** knowledge, technology, potential innovation, and human skills valuable to human prosperity.

25. **International Institute for Sustainable Development:** an independent, nonprofit organization founded to promote human development by sustainable means. The organization reports on international negotiations, encourages innovation and communication, and attempts to engage citizens, businesses, and policymakers in issues surrounding sustainability.

26. **Kyoto Protocol:** an international treaty adopted in Kyoto, Japan, in 1997 that aimed to reduce carbon dioxide (CO_2) emissions and slow climate change.

27. **Millennium Development Goals:** eight international development goals introduced in 2000 by the UN Millennium Summit, originally scheduled to be achieved by 2015. These goals include commitments to eradicate poverty, provide primary education, protect women's rights, decrease child mortality rates, improve maternal health, combat diseases such as HIV/AIDS, maintain environmental sustainability, and foster international partnerships globally.

28. **Monitoring and evaluation:** a process in project management in which an activity is constantly checked to see if it is effective, and information about it is continuously updated.

29. **Multilateralism:** a concept in international relations developed by Miles Kahler, promoting international governance by multiple states.

30. **Natural capital:** the supply of natural resources, including geology, air, water, soil, and living organisms. It can be used for economic activity.

31. **Neoliberalism:** a school of economic thought that emerged in the nineteenth century and promotes policies of economic liberalization and free trade.

32. **Nongovernmental organization (NGO):** a nonprofit voluntary group that is organized at local, national, or international level.

33. **Norwegian Labor Party:** a social-democratic political party founded in 1887 and still active in the political arena of Norway promoting socialists values.

34. **Palme Commission:** the Commission on Disarmament and Security, chaired by Swedish politician Olaf Palme, that produced the report *Common Security* in 1982.

35. **Paris Conference:** the United Nations Climate Change Conference held in Paris in December 2015; it concluded with the proposal of a global agreement to reduce climate change (the Paris Agreement).

36. **Policy:** a course of action followed by a government, a political party, or a company.

37. **Rio+20 Conference:** the United Nations Conference on Sustainable Development (UNCSD) held in Rio de Janeiro in 2012 to follow-up on the activities proposed by the 1992 Earth Summit.

38. **Stockholm Conference:** also known as the United Nations Conference on the Human Environment, this meeting was held in Stockholm, on June 5–16, 1972. The conference was an important step towards formulating the sustainable development concept, as it elaborated principles of actions to balance environmental challenges and socioeconomic development, known as the Stockholm Declaration. The idea to establish the United Nations Environment Program (UNEP) was also put forward at the conference.

39. **Soviet Union:** the Union of Soviet Socialist Republics (USSR) was a socialist state that existed between 1922 and 1991, centered on Russia. It was a one-party state, governed by the Communist Party, with Moscow as its capital.

40. **Special envoy:** a person chosen by the UN secretary-general to act as a representative and adviser on a particular matter, such as human rights or climate change. The position is honorary and unpaid.

41. **Strong sustainability:** the idea supported by ecological environmentalists that human capital and natural capital may be complementary, but they are not interchangeable.

42. **Sustainability:** the main principle of responsible use and management of resources, presented in *Our Common Future*: a development model that "meets the needs of the present without compromising the ability of future generations to meet their own needs."

43. **Sustainable development:** the various policies and strategies designed to improve the lives of people while ensuring that future generations will be able to meet their needs.

44. **Sustainable Development Goals (SDGs):** a set of 17 goals to improve the condition of living globally. The idea of SDGs was discussed in 2012 at the Rio+20 Conference and accepted by the UN General Assembly in 2014.

45. **Think tank:** a group of experts providing advice and ideas on specific political or economic problems.

46. **United Nations (UN):** an intergovernmental organization created in 1945 by 51 founding member states to promote international cooperation and global peace.

47. **United Nations Environment Program (UNEP):** a UN agency specializing in environmental activities. It was created in 1972 on the initiative of Maurice Strong. Its headquarters are in Nairobi, Kenya.

48. **Weak sustainability:** an idea supported by environmental economics that contends that human capital and natural capital are interchangeable.

49. **World Commission on Environment and Development:** the original title of the Brundtland Commission, convened by the United Nations with the aim of fostering cooperation between nations to pursue sustainable development.

50. **World Health Organization (WHO):** one of the United Nations agencies specializing in global public health. Its headquarters are based in Geneva.

51. **Yugoslavia:** a European country that existed from 1918 until 1991. Its territory included what is today Bosnia and Herzegovina, Croatia, Macedonia, Montenegro, Slovenia, and Serbia.

PEOPLE MENTIONED IN THE TEXT

1. **Lucien Bouchard (b. 1938)** is a former minister of the environment in the Canadian government. A politician and diplomat, he was premier of the Canadian region of Quebec between 1996 and 2001.

2. **Boutros Boutros-Ghali (1922–2016)** was an Egyptian politician who served as secretary-general of the United Nations from 1992 to 1996.

3. **Willy Brandt (1913–92)** was chancellor of the Federal Republic of Germany from 1969 to 1974. In 1980 he chaired an Independent Commission that produced the *Brandt Report* on global economic development.

4. **Lester R. Brown (b. 1934)** is an American environmentalist and the author of several books discussing major challenges of the contemporary world and different scenarios of the human future. He is a founder of the think tank Worldwatch Institute and has written or coauthored more than 50 books, including *Man, Land and Food* (1963).

5. **Hillary Clinton (b. 1947)** is an American politician who served as secretary of state under President Barack Obama. She is a candidate for United States president in 2016.

6. **Herman Daly (b. 1938)** is an American professor of economics at the School of Public Policy of the University of Maryland who specializes in environmental economics.

7. **Jacques Delors (b. 1925)** is a French economist who served as the president of the European Commission between 1985 and 1995.

8. **Nitin Desai (b. 1941)** is an Indian economist and, as a member of the Brundtland Commission, an author of *Our Common Future*. He also served as under-secretary-general for economic and social affairs of the United Nations from 1992 to 2003.

9. **Arturo Escobar (b. 1952)** is a Columbian American anthropologist who studies social movements and international development. He is the Kenan Distinguished Professor of Anthropology at the University of North Carolina at Chapel Hill.

10. **Matthias Finger (b. 1955)** is the dean of the School of Continuing Education in the College of Management of Technology at the Swiss Federal Institute of

Technology in Lausanne.

11. **Volker Hauff (b. 1940)** is a German politician; he is a member of the Social Democratic Party.

12. **Wes Jackson (b. 1936)** is one of the founders and leaders of sustainable agriculture movements. He has studied the relationship between human agricultural activity and ecosystems, and argued for more careful use of technology and nonrenewable resources.

13. **Ban Ki-moon (b. 1944)** is a former South Korean minister of foreign affairs and trade, and the current secretary-general of the United Nations.

14. **Drago Kos (b. 1961)** is a Slovenian lawyer, journalist, and former police officer. He was the first president of the OECD Working Group on Bribery.

15. **Thomas Robert Malthus (1766–1834)** was an English cleric and scholar who discussed the connection between famine and population growth, known as a Malthusian catastrophe. He published his most significant work, *An Essay on the Principle of Population*, in 1798.

16. **Nelson Rolihlahla Mandela (1918–2013)** was a South African human rights activist who served as the president of South Africa from 1994 to 1999.

17. **Francois Maurice Adrien Marie Mitterand (1916–96)** was a leader of the Socialist Party of France who served as the president of France from 1981 to 1995.

18. **Saburo Okita (1914–93)** was a Japanese economist and a minister of foreign affairs (1979–80) who chaired research on Japanese economic growth.

19. **Sven Olaf Joachim Palme (1927–86)** was a Swedish prime minister. He was chairman of the Commission on Disarmament and Security, also known as the Palme Commission, that produced the report *Common Security* in 1982.

20. **Javier Perez de Cuellar (1920–2020)** was a former prime minister of Peru and the fifth secretary-general of the United Nations. He commissioned Gro Brundtland to chair the World Commission on Environment and Development and write *Our Common Future* (the *Brundtland Report*).

21. **Michael E. Porter (b. 1947)** is an American economist who writes on corporate social responsibility and green business. He is the Bishop William Lawrence University Professor at the Institute for Strategy and Competitiveness at the Harvard Business School. He developed the "five forces analysis," a framework for analyzing the level of competition within an industry and business development.
22. **Jeffrey D. Sachs (b. 1954)** is an American economist known as an expert in development and poverty, and a special adviser to the secretary-general of the United Nations on sustainable development. He is the author of *The End of Poverty* (2005), *CommonWealth* (2008), and *The Price of Civilization* (2011).
23. **Vandana Shiva (b. 1952)** is an Indian environmentalist, feminist, and international activist. She leads campaigns against genetic engineering and works closely with grassroots organizations in developing countries.
24. **Maurice Strong (1929–2015)** was a Canadian businessman, the first director of UNEP, and one of the Brundtland commissioners who wrote *Our Common Future*.
25. **Desmond Mpilo Tutu (b. 1931)** is a retired bishop and a human rights activist who fought against racial discrimination in South Africa.
26. **Shiv Visvanathan** is an Indian public intellectual and social scientist.
27. **Boris Yeltsin (1931–2007)** was the first president of the Russian Federation.

WORKS CITED

1. Brown, Lester R. *Man, Land and Food: Looking Ahead at World Food Needs*. Washington, DC: US Department of Agriculture, 1963.
2. Brundtland, Arne Olav. *Gift med Gro*. Oslo: Oslo University, 1996.
3. ———. *Fortsatt Gift med Gro*. Oslo: Oslo University, 2003.
4. Brundtland, Gro Harlem. "Global Change and Our Common Future: The Benjamin Franklin Lecture." In *Global Change and Our Common Future*, edited by R. S. DeFries and T. F. Malone. Washington, DC: National Academy Press, 1989.
5. ———. *Interview with Peter Ocskay*. Baltic University, Uppsala, 1997. Accessed February 1, 2016. https://www.youtube.com/watch?v=ogrcy8AY95I.
6. ———. *Mitt Liv: 1939–1986*. Oslo: Gylenhal, 1997.
7. ———. *Dramatiske ar 1986–1996*. Oslo: Gyldedal, 1998.
8. ———. "Healthy People, Healthy Planet." The Annual Lecture in the Business and the Environment Program. London, March 15, 2001. Accessed February 1, 2016. http://www.cisl.cam.ac.uk/publications/archive-publications/brundtland-paper.
9. ———. *Madam Prime Minister: A Life in Power and Politics*. New York: Farrar, Straus and Giroux, 2005.
10. ———. *Interview with the Uongozi Institute*. Oslo, July 5, 2013. Accessed February 1, 2016. https://www.youtube.com/watch?v=8POtjDtH6io.
11. ———. "ARA Lecture." Vienna Technical University. November 18, 2013. Accessed February 1, 2016. https://www.youtube.com/watch?v=X7Z2o8tZZoE.
12. ———. *Opening Speech to the 2015 Nobel Peace Prize Forum*. Oslo, March 10, 2015. Accessed February 1, 2016. https://www.youtube.com/watch?v=LBRmSjsnVGs.
13. Brundtland, Gro Harlem, and Jeffrey D. Sachs. "Macroeconomics and Health: Investing in Health for Economic Development." Geneva: World Health Organization, 2001.
14. Chatterjee, Pratap, and Matthias Finger. *The Earth Brokers: Power, Politics and World Development*. London: Routledge, 2013.

15. Daly, Herman. *Ecological Economics and Sustainable Development: Selected Essays*. Cheltenham: Edward Elgar, 2007.
16. Desai, Nitin. "Symposium: The Road from Johannesburg." Keynote Address. Georgetown: *Environmental Law Review*, 2003.
17. Dodds, Felix, and Michael Strauss with Maurice F. Strong. *Only One Earth: The Long Road via Rio to Sustainable Development*. London: Routledge, 2012.
18. Drexhage, John, and Deborah Murphy. *Sustainable Development: From Brundtland to Rio 2012*. New York: United Nations, 2010.
19. Elliott, Jennifer. *An Introduction to Sustainable Development: The Developing World*. London: Routledge, 2000.
20. Escobar, Arturo. "Culture Sits in Places: Reflections on Globalism and Subaltern Strategies of Localization." *Political Geography* 20 (2001): 139–74.
21. ———. "Alternatives to Development." Interview with Rob Hopkins, Venice, September 28, 2012. Accessed February 1, 2016. http://transitionculture.org/2012/09/28/alternatives-to-development-an-interview-with-arturo-escobar/.
22. Food and Agriculture Organization. *How to Feed the World in 2050*. Rome: Food and Agriculture Organization, 2009.
23. Finger, Matthias. "Politics of the UNCED Process." In *Global Ecology: A New Arena of Political Conflict*, edited by Wolfgang Sachs. London: Zed Books, 1993.
24. Global Alliance for the Rights of Nature. "Universal Declaration of the Rights of Mother Earth." Cochabamba, Bolivia, April 22, 2010. Accessed December 14, 2015. http://therightsofnature.org/universal-declaration/.
25. Gowdy, John M., and Marsha Walton. "Sustainability Concepts in Ecological Economics." *Economics Interactions with Other Disciplines, Volume 2. Encyclopedia of Life Support Systems*. Paris: UNESCO/Eolss, 2008.
26. Hauff, Volker. "Brundtland Report: A 20 Years Update." Keynote speech, European Sustainability, Berlin, June 3, 2007. Accessed February 2, 2016. http://www.nachhaltigkeitsrat.de/uploads/media/ESB07_Keynote_speech_

Hauff_07-06-04_01.pdf.

27. Jackson, Wes. *New Roots for Agriculture*. Lincoln: University of Nebraska Press, 1985.

28. Kos, Drago. "Sustainable Development: Implementing Utopia?" *SOCIOLOGIJA* 54, no. 1 (2012).

29. Leape, Jim. "It's Happening, but Not in Rio." *New York Times*, June 24, 2012. Accessed February 1, 2016. http://www.nytimes.com/2012/06/25/opinion/action-is-happening-but-not-in-rio.html?_r=0.

30. Malthus, Thomas R. *An Essay on the Principle of Population*. London: J. Johnson, 1798.

31. Meadows, Donella H., Dennis L. Meadows, Jorgen Randers, and William W. Behrens III. *The Limits to Growth: A Report for the Club of Rome's Project on the Predicament of Mankind*. New York: Universe Books, 1974.

32. Norgard, Jorgen, John Peet, and Kristin Ragnarsdottir. "The History of the Limits to Growth." *Solutions* 2, no. 1 (2010): 59–63.

33. O'Neill, D. W., R. Dietz, and N. Jones. "Enough Is Enough: Ideas for a Sustainable Economy in a World of Finite Resources." *The Report of the Steady State Economy Conference*. Leeds: CASSE and Economic Justice for All, 2010.

34. O Tuathail, Gearoid, Simon Dalby and Paul Routledge (eds). *The Geopolitics Reader*. London: Routledge, 1998.

35. Parenivel, Mauree. "Sustainability and Sustainable Development." *Le Mauricien*, August 17, 2011. Accessed February 1, 2016. http://www.lemauricien.com/article/maurice-ile-durable-sustainability-and-sustainable-development.

36. Pearce, David, and Giles Atkinson. "The Concept of Sustainable Development: An Evaluation of its Usefulness Ten Years after Brundtland." *Swiss Journal of Economics and Statistics* 134, no. 3 (1998).

37. Pelenc, Jerome. "Weak Sustainability versus Strong Sustainability." Brief for *GSDR*. Louvain: UCL, 2015.

38. Pereira, Winin, and Jeremy Seabrook. *Asking the Earth: Farms, Forests and*

Survival in India. Sterling, VA: Earthscan, 1990.

39. Pile, Ben. "Wishing Greenpeace an Unhappy Birthday." *Spiked*, September 12, 2011. Accessed February 2, 2016. http://www.spiked-online.com/newsite/article/11068#.VrCGxPkS-Uk.

40. Porter, Michael E., and Claas van der Linde. "Green and Competitive: Ending the Stalemate." *Harvard Business Review* 73, no. 5 (1995).

41. Shadish, William R., Thomas D. Cook, and Laura C. Leviton. *Foundations of Program Evaluations: Theories of Practice*. London: Sage, 1991.

42. Shiva, Vandana. "The Greening of Global Reach." In *Global Ecology: A New Arena of Political Conflict*, edited by Wolfgang Sachs. London: Zed Books, 1993.

43. Starke, Linda. *Signs of Hope: Working towards Our Common Future*. Oxford: Oxford University Press, 1990.

44. Sterner, Thomas. "The Paris Climate Change Conference Needs to Be More Ambitious." *Economist*, November 18, 2015. Accessed February 2, 2016. http://www.economist.com/blogs/freeexchange/2015/11/cutting-carbon-emissions.

45. Strong, Maurice F. "The 1992 Earth Summit: An Inside View." Interview with Philip Shabecoff, Quebec, 1999. Accessed February 1, 2016. http://www.mauricestrong.net/index.php/earth-summit-strong.

46. ———. *Where on Earth Are We Going?* Toronto: Vintage Canada, 2001.

47. ———. "Statement by Maurice F. Strong Delivered to the Special United Nations General Assembly Event on Rio+20." New York, October 25, 2011. Accessed February 1, 2016. http://www.unep.org/environmentalgovernance/PerspectivesonRIO20/MauriceFStrong/tabid/55711/Default.aspx.

48. Turner, Graham. *A Comparison of Limits to Growth with Thirty Years of Reality*. CSIRO Working Paper. Canberra: CSIRO, 2008.

49. Tutu, Desmond Mpilo. "Is Sustainable Development a Luxury We Can't Afford?" Interview with the Elders, Cape Town, May 12, 2012. Accessed February 1, 2016. https://www.youtube.com/watch?v=ArHey8SVJQE.

50. United Nations. *Biography of Dr Gro Harlem Brundtland*. Geneva: UN, 2014.
51. Victor, P. A., J. E. Hanna, and A. Kubursi. "How Strong is Weak Sustainability?" *Economie Appliquée* 48, no. 2 (1995): 75–94.
52. Visvanathan, Shiv. "Mrs Brundtland's Disenchanted Cosmos." *Alternatives* 16 (1991): 378–81.
53. Wasdell, David. *Studies in Global Dynamics No. 7—Brundtland and Beyond: Towards a Global Process*. London: Urchin: 1987.
54. Wilsford, David. *Political Leaders of Contemporary Western Europe: A Biographical Dictionary*. Westport, CT: Greenwood Press, 1995.
55. World Commission on Environment and Development. *Report of the World Commission on Environment and Development: Our Common Future*. August 4, 1987. Accessed February 1, 2016. http://www.un-documents.net/wced-ocf.htm.
56. ———. *Our Common Future*. Oxford: Oxford University Press, 1989.
57. World Health Organization. "Dr Gro Harlem Brundtland, Director-General." Geneva: WHO, 1998. Accessed March 9, 2016. http://www.who.int/dg/brundtland/bruntland/en/.

原书作者简介

格罗·哈莱姆·布伦特兰于1939年出生于挪威首都奥斯陆，在成为政治家之前曾接受医生培训。她于1974年至1979年期间担任环境大臣，1981年成为挪威首相——第一位担任首相职位的女性。1983年，她被聘为联合国世界环境与发展委员会主席，并负责制定《我们共同的未来》，该报告于1987年出版。布伦特兰于1986—1989年及1990—1996年连续担任两届首相。现在，她是一位杰出的政治人物，并担任长老会（由世界领导人组成的极具影响力的组织）的副主席。

本书作者简介

克谢尼娅·格拉西莫娃博士，剑桥大学博士，现任剑桥大学发展研究中心客座讲师。

世界名著中的批判性思维

《世界思想宝库钥匙丛书》致力于深入浅出地阐释全世界著名思想家的观点，不论是谁、在何处都能了解到，从而推进批判性思维发展。

《世界思想宝库钥匙丛书》与世界顶尖大学的一流学者合作，为一系列学科中最有影响的著作推出新的分析文本，介绍其观点和影响。在这一不断扩展的系列中，每种选入的著作都代表了历经时间考验的思想典范。通过为这些著作提供必要背景、揭示原作者的学术渊源以及说明这些著作所产生的影响，本系列图书希望让读者以新视角看待这些划时代的经典之作。读者应学会思考、运用并挑战这些著作中的观点，而不是简单接受它们。

ABOUT THE AUTHOR OF THE ORIGINAL WORK

Gro Harlem Brundtland was born in Oslo, Norway, in 1939, and trained as a doctor before becoming a politician. She served as minister for environmental affairs from 1974 to 1979 and then, in 1981, became prime minister of Norway—the first woman to do so. In 1983 she was recruited to chair the United Nations World Commission on Environment and Development and to oversee its report, *Our Common Future*, which was published in 1987. Brundtland served two more terms as prime minister, from 1986 to 1989 and from 1990 to 1996. Today, she is a prominent political figure and deputy chair of the Elders, an influential group of world leaders.

ABOUT THE AUTHOR OF THE ANALYSIS

Dr Ksenia Gerasimova holds a PhD from the University of Cambridge, where she is currently an Affiliated Lecturer at the Centre for Development Studies.

ABOUT MACAT
GREAT WORKS FOR CRITICAL THINKING

Macat is focused on making the ideas of the world's great thinkers accessible and comprehensible to everybody, everywhere, in ways that promote the development of enhanced critical thinking skills.

It works with leading academics from the world's top universities to produce new analyses that focus on the ideas and the impact of the most influential works ever written across a wide variety of academic disciplines. Each of the works that sit at the heart of its growing library is an enduring example of great thinking. But by setting them in context — and looking at the influences that shaped their authors, as well as the responses they provoked — Macat encourages readers to look at these classics and game-changers with fresh eyes. Readers learn to think, engage and challenge their ideas, rather than simply accepting them.

批判性思维与《我们共同的未来》

首要批判性思维技巧：解决问题
次要批判性思维技巧：评估

  《我们共同的未来》是由挪威前首相格罗·布伦特兰领导的一个联合国委员会合作撰写并于 1987 年出版的一份报告，也称《布伦特兰报告》。为了提供有效的解决问题的方案，该报告提出了很多富有创造性的问题，是批判性思维技巧得以运用的一个杰出示例。

  布伦特兰报告提出这样一个问题，我们如何能够既为后代保护好我们当前生活的世界，同时又能促进当今的经济和社会发展。该报告提供的解决方案是"可持续发展"，并将其定义为人类"既满足当前的需要，又不损害后代满足其自身需要"的能力。该报告得出的主要结论——我们需要制定长期战略来管理地球的自然资源——得到普遍认可，由此"可持续性"这个术语被纳入国际政治的日常语言中。

  解决可持续发展问题成为一个热门话题，并催生了一门新的学科——环境经济学。通过提出正确的问题——批判性评估问题的答案——《我们共同的未来》强调国际合作的重要性，为确保可持续发展提供了解决方案。

CRITICAL THINKING AND *OUR COMMON FUTURE*

- Primary critical thinking skill: PROBLEM-SOLVING
- Secondary critical thinking skill: EVALUATION

Our Common Future is a joint work produced in 1987 by a United Nations commission headed by former Norwegian Prime Minister, Gro Brundtland. Also known as *The Brundtland Report*, it is a striking example of the critical thinking skill of asking productive questions in order to produce valid solutions to a problem.

Brundtland's report asks how we can protect the world we live in for future generations, while at the same time stimulating economic and social development now. The solution the work proposes is 'sustainable development', defined as humanity's ability 'to ensure that it meets the needs of the present without compromising the ability of future generations to meet their own needs.' The key conclusion the report came to — that we need long-term strategies to manage the earth's natural resources — proved to be so universally welcomed it introduced the term 'sustainability' into the everyday language of international politics.

Solving the problem of workable sustainable development became a hot topic, leading to the birth of a new academic discipline, environmental economics. By asking the right questions — and critically evaluating the answers to those questions — *Our Common Future* offered a solution to the problem of ensuring sustainable development by highlighting the critical importance of international cooperation.

《世界思想宝库钥匙丛书》简介

《世界思想宝库钥匙丛书》致力于为一系列在各领域产生重大影响的人文社科类经典著作提供独特的学术探讨。每一本读物都不仅仅是原经典著作的内容摘要,而是介绍并深入研究原经典著作的学术渊源、主要观点和历史影响。这一丛书的目的是提供一套学习资料,以促进读者掌握批判性思维,从而更全面、深刻地去理解重要思想。

每一本读物分为3个部分:学术渊源、学术思想和学术影响,每个部分下有4个小节。这些章节旨在从各个方面研究原经典著作及其反响。

由于独特的体例,每一本读物不但易于阅读,而且另有一项优点:所有读物的编排体例相同,读者在进行某个知识层面的调查或研究时可交叉参阅多本该丛书中的相关读物,从而开启跨领域研究的路径。

为了方便阅读,每本读物最后还列出了术语表和人名表(在书中则以星号﹡标记),此外还有参考文献。

《世界思想宝库钥匙丛书》与剑桥大学合作,理清了批判性思维的要点,即如何通过6种技能来进行有效思考。其中3种技能让我们能够理解问题,另3种技能让我们有能力解决问题。这6种技能合称为"批判性思维PACIER模式",它们是:

分析:了解如何建立一个观点;
评估:研究一个观点的优点和缺点;
阐释:对意义所产生的问题加以理解;
创造性思维:提出新的见解,发现新的联系;
解决问题:提出切实有效的解决办法;
理性化思维:创建有说服力的观点。

THE MACAT LIBRARY

The Macat Library is a series of unique academic explorations of seminal works in the humanities and social sciences — books and papers that have had a significant and widely recognised impact on their disciplines. It has been created to serve as much more than just a summary of what lies between the covers of a great book. It illuminates and explores the influences on, ideas of, and impact of that book. Our goal is to offer a learning resource that encourages critical thinking and fosters a better, deeper understanding of important ideas.

Each publication is divided into three Sections: Influences, Ideas, and Impact. Each Section has four Modules. These explore every important facet of the work, and the responses to it.

This Section-Module structure makes a Macat Library book easy to use, but it has another important feature. Because each Macat book is written to the same format, it is possible (and encouraged!) to cross-reference multiple Macat books along the same lines of inquiry or research. This allows the reader to open up interesting interdisciplinary pathways.

To further aid your reading, lists of glossary terms and people mentioned are included at the end of this book (these are indicated by an asterisk [*] throughout) — as well as a list of works cited.

Macat has worked with the University of Cambridge to identify the elements of critical thinking and understand the ways in which six different skills combine to enable effective thinking.

Three allow us to fully understand a problem; three more give us the tools to solve it. Together, these six skills make up the PACIER model of critical thinking. They are:

ANALYSIS — understanding how an argument is built
EVALUATION — exploring the strengths and weaknesses of an argument
INTERPRETATION — understanding issues of meaning
CREATIVE THINKING — coming up with new ideas and fresh connections
PROBLEM-SOLVING — producing strong solutions
REASONING — creating strong arguments

"《世界思想宝库钥匙丛书》提供了独一无二的跨学科学习和研究工具。它介绍那些革新了各自学科研究的经典著作,还邀请全世界一流专家和教育机构进行严谨的分析,为每位读者打开世界顶级教育的大门。"

—— 安德烈亚斯·施莱歇尔,
经济合作与发展组织教育与技能司司长

"《世界思想宝库钥匙丛书》直面大学教育的巨大挑战……他们组建了一支精干而活跃的学者队伍,来推出在研究广度上颇具新意的教学材料。"

—— 布罗尔斯教授、勋爵,剑桥大学前校长

"《世界思想宝库钥匙丛书》的愿景令人赞叹。它通过分析和阐释那些曾深刻影响人类思想以及社会、经济发展的经典文本,提供了新的学习方法。它推动批判性思维,这对于任何社会和经济体来说都是至关重要的。这就是未来的学习方法。"

—— 查尔斯·克拉克阁下,英国前教育大臣

"对于那些影响了各自领域的著作,《世界思想宝库钥匙丛书》能让人们立即了解到围绕那些著作展开的评论性言论,这让该系列图书成为在这些领域从事研究的师生们不可或缺的资源。"

—— 威廉·特朗佐教授,加利福尼亚大学圣地亚哥分校

"Macat offers an amazing first-of-its-kind tool for interdisciplinary learning and research. Its focus on works that transformed their disciplines and its rigorous approach, drawing on the world's leading experts and educational institutions, opens up a world-class education to anyone."

—— Andreas Schleicher, Director for Education and Skills, Organisation for Economic Co-operation and Development

"Macat is taking on some of the major challenges in university education... They have drawn together a strong team of active academics who are producing teaching materials that are novel in the breadth of their approach."

—— Prof Lord Broers, former Vice-Chancellor of the University of Cambridge

"The Macat vision is exceptionally exciting. It focuses upon new modes of learning which analyse and explain seminal texts which have profoundly influenced world thinking and so social and economic development. It promotes the kind of critical thinking which is essential for any society and economy. This is the learning of the future."

—— Rt Hon Charles Clarke, former UK Secretary of State for Education

"The Macat analyses provide immediate access to the critical conversation surrounding the books that have shaped their respective discipline, which will make them an invaluable resource to all of those, students and teachers, working in the field."

—— Prof William Tronzo, University of California at San Diego

The Macat Library
世界思想宝库钥匙丛书

TITLE	中文书名	类别
An Analysis of Arjun Appadurai's *Modernity at Large: Cultural Dimensions of Globalization*	解析阿尔君·阿帕杜莱《消失的现代性：全球化的文化维度》	人类学
An Analysis of Claude Lévi-Strauss's *Structural Anthropology*	解析克劳德·列维−斯特劳斯《结构人类学》	人类学
An Analysis of Marcel Mauss's *The Gift*	解析马塞尔·莫斯《礼物》	人类学
An Analysis of Jared M. Diamond's *Guns, Germs, and Steel: The Fate of Human Societies*	解析贾雷德·M.戴蒙德《枪炮、病菌与钢铁：人类社会的命运》	人类学
An Analysis of Clifford Geertz's *The Interpretation of Cultures*	解析克利福德·格尔茨《文化的解释》	人类学
An Analysis of Philippe Ariès's *Centuries of Childhood: A Social History of Family Life*	解析菲力浦·阿利埃斯《儿童的世纪：旧制度下的儿童和家庭生活》	人类学
An Analysis of W. Chan Kim & Renée Mauborgne's *Blue Ocean Strategy*	解析金伟灿/勒妮·莫博涅《蓝海战略》	商业
An Analysis of John P. Kotter's *Leading Change*	解析约翰·P.科特《领导变革》	商业
An Analysis of Michael E. Porter's *Competitive Strategy: Techniques for Analyzing Industries and Competitors*	解析迈克尔·E.波特《竞争战略：分析产业和竞争对手的技术》	商业
An Analysis of Jean Lave & Etienne Wenger's *Situated Learning: Legitimate Peripheral Participation*	解析琼·莱夫/艾蒂纳·温格《情境学习：合法的边缘性参与》	商业
An Analysis of Douglas McGregor's *The Human Side of Enterprise*	解析道格拉斯·麦格雷戈《企业的人性面》	商业
An Analysis of Milton Friedman's *Capitalism and Freedom*	解析米尔顿·弗里德曼《资本主义与自由》	商业
An Analysis of Ludwig von Mises's *The Theory of Money and Credit*	解析路德维希·冯·米塞斯《货币和信用理论》	经济学
An Analysis of Adam Smith's *The Wealth of Nations*	解析亚当·斯密《国富论》	经济学
An Analysis of Thomas Piketty's *Capital in the Twenty-First Century*	解析托马斯·皮凯蒂《21世纪资本论》	经济学
An Analysis of Nassim Nicholas Taleb's *The Black Swan: The Impact of the Highly Improbable*	解析纳西姆·尼古拉斯·塔勒布《黑天鹅：如何应对不可预知的未来》	经济学
An Analysis of Ha-Joon Chang's *Kicking Away the Ladder*	解析张夏准《富国陷阱：发达国家为何踢开梯子》	经济学
An Analysis of Thomas Robert Malthus's *An Essay on the Principle of Population*	解析托马斯·罗伯特·马尔萨斯《人口论》	经济学

An Analysis of John Maynard Keynes's *The General Theory of Employment, Interest and Money*	解析约翰·梅纳德·凯恩斯《就业、利息和货币通论》	经济学
An Analysis of Milton Friedman's *The Role of Monetary Policy*	解析米尔顿·弗里德曼《货币政策的作用》	经济学
An Analysis of Burton G. Malkiel's *A Random Walk Down Wall Street*	解析伯顿·G.马尔基尔《漫步华尔街》	经济学
An Analysis of Friedrich A. Hayek's *The Road to Serfdom*	解析弗里德里希·A.哈耶克《通往奴役之路》	经济学
An Analysis of Charles P. Kindleberger's *Manias, Panics, and Crashes: A History of Financial Crises*	解析查尔斯·P.金德尔伯格《疯狂、惊恐和崩溃：金融危机史》	经济学
An Analysis of Amartya Sen's *Development as Freedom*	解析阿马蒂亚·森《以自由看待发展》	经济学
An Analysis of Rachel Carson's *Silent Spring*	解析蕾切尔·卡森《寂静的春天》	地理学
An Analysis of Charles Darwin's *On the Origin of Species: by Means of Natural Selection, or The Preservation of Favoured Races in the Struggle for Life*	解析查尔斯·达尔文《物种起源》	地理学
An Analysis of World Commission on Environment and Development's *The Brundtland Report: Our Common Future*	解析世界环境与发展委员会《布伦特兰报告：我们共同的未来》	地理学
An Analysis of James E. Lovelock's *Gaia: A New Look at Life on Earth*	解析詹姆斯·E.拉伍洛克《盖娅：地球生命的新视野》	地理学
An Analysis of Paul Kennedy's *The Rise and Fall of the Great Powers: Economic Change and Military Conflict from 1500–2000*	解析保罗·肯尼迪《大国的兴衰：1500—2000年的经济变革与军事冲突》	历史
An Analysis of Janet L. Abu-Lughod's *Before European Hegemony: The World System A. D. 1250–1350*	解析珍妮特·L.阿布-卢格霍德《欧洲霸权之前：1250—1350年的世界体系》	历史
An Analysis of Alfred W. Crosby's *The Columbian Exchange: Biological and Cultural Consequences of 1492*	解析艾尔弗雷德·W.克罗斯比《哥伦布大交换：1492以后的生物影响和文化冲击》	历史
An Analysis of Tony Judt's *Postwar: A History of Europe since 1945*	解析托尼·朱特《战后欧洲史》	历史
An Analysis of Richard J. Evans's *In Defence of History*	解析理查德·J.艾文斯《捍卫历史》	历史
An Analysis of Eric Hobsbawm's *The Age of Revolution: Europe 1789–1848*	解析艾瑞克·霍布斯鲍姆《革命的年代：欧洲1789—1848年》	历史

An Analysis of Roland Barthes's *Mythologies*	解析罗兰·巴特《神话学》	文学与批判理论
An Analysis of Simone de Beauvoir's *The Second Sex*	解析西蒙娜·德·波伏娃《第二性》	文学与批判理论
An Analysis of Edward W. Said's *Orientalism*	解析爱德华·W. 萨义德《东方主义》	文学与批判理论
An Analysis of Virginia Woolf's *A Room of One's Own*	解析弗吉尼亚·伍尔芙《一间自己的房间》	文学与批判理论
An Analysis of Judith Butler's *Gender Trouble*	解析朱迪斯·巴特勒《性别麻烦》	文学与批判理论
An Analysis of Ferdinand de Saussure's *Course in General Linguistics*	解析费尔迪南·德·索绪尔《普通语言学教程》	文学与批判理论
An Analysis of Susan Sontag's *On Photography*	解析苏珊·桑塔格《论摄影》	文学与批判理论
An Analysis of Walter Benjamin's *The Work of Art in the Age of Mechanical Reproduction*	解析瓦尔特·本雅明《机械复制时代的艺术作品》	文学与批判理论
An Analysis of W. E. B. Du Bois's *The Souls of Black Folk*	解析W.E.B. 杜波依斯《黑人的灵魂》	文学与批判理论
An Analysis of Plato's *The Republic*	解析柏拉图《理想国》	哲学
An Analysis of Plato's *Symposium*	解析柏拉图《会饮篇》	哲学
An Analysis of Aristotle's *Metaphysics*	解析亚里士多德《形而上学》	哲学
An Analysis of Aristotle's *Nicomachean Ethics*	解析亚里士多德《尼各马可伦理学》	哲学
An Analysis of Immanuel Kant's *Critique of Pure Reason*	解析伊曼努尔·康德《纯粹理性批判》	哲学
An Analysis of Ludwig Wittgenstein's *Philosophical Investigations*	解析路德维希·维特根斯坦《哲学研究》	哲学
An Analysis of G. W. F. Hegel's *Phenomenology of Spirit*	解析G.W.F. 黑格尔《精神现象学》	哲学
An Analysis of Baruch Spinoza's *Ethics*	解析巴鲁赫·斯宾诺莎《伦理学》	哲学
An Analysis of Hannah Arendt's *The Human Condition*	解析汉娜·阿伦特《人的境况》	哲学
An Analysis of G. E. M. Anscombe's *Modern Moral Philosophy*	解析G. E. M. 安斯康姆《现代道德哲学》	哲学
An Analysis of David Hume's *An Enquiry Concerning Human Understanding*	解析大卫·休谟《人类理解研究》	哲学

An Analysis of Søren Kierkegaard's *Fear and Trembling*	解析索伦·克尔凯郭尔《恐惧与战栗》	哲学
An Analysis of René Descartes's *Meditations on First Philosophy*	解析勒内·笛卡尔《第一哲学沉思录》	哲学
An Analysis of Friedrich Nietzsche's *On the Genealogy of Morality*	解析弗里德里希·尼采《论道德的谱系》	哲学
An Analysis of Gilbert Ryle's *The Concept of Mind*	解析吉尔伯特·赖尔《心的概念》	哲学
An Analysis of Thomas Kuhn's *The Structure of Scientific Revolutions*	解析托马斯·库恩《科学革命的结构》	哲学
An Analysis of John Stuart Mill's *Utilitarianism*	解析约翰·斯图亚特·穆勒《功利主义》	哲学
An Analysis of Aristotle's *Politics*	解析亚里士多德《政治学》	政治学
An Analysis of Niccolò Machiavelli's *The Prince*	解析尼科洛·马基雅维利《君主论》	政治学
An Analysis of Karl Marx's *Capital*	解析卡尔·马克思《资本论》	政治学
An Analysis of Benedict Anderson's *Imagined Communities*	解析本尼迪克特·安德森《想象的共同体》	政治学
An Analysis of Samuel P. Huntington's *The Clash of Civilizations and the Remaking of World Order*	解析塞缪尔·P.亨廷顿《文明的冲突与世界秩序的重建》	政治学
An Analysis of Alexis de Tocqueville's *Democracy in America*	解析阿列克西·德·托克维尔《论美国的民主》	政治学
An Analysis of John A. Hobson's *Imperialism: A Study*	解析约翰·A.霍布森《帝国主义》	政治学
An Analysis of Thomas Paine's *Common Sense*	解析托马斯·潘恩《常识》	政治学
An Analysis of John Rawls's *A Theory of Justice*	解析约翰·罗尔斯《正义论》	政治学
An Analysis of Francis Fukuyama's *The End of History and the Last Man*	解析弗朗西斯·福山《历史的终结与最后的人》	政治学
An Analysis of John Locke's *Two Treatises of Government*	解析约翰·洛克《政府论》	政治学
An Analysis of Sun Tzu's *The Art of War*	解析孙武《孙子兵法》	政治学
An Analysis of Henry Kissinger's *World Order: Reflections on the Character of Nations and the Course of History*	解析亨利·基辛格《世界秩序》	政治学
An Analysis of Jean-Jacques Rousseau's *The Social Contract*	解析让-雅克·卢梭《社会契约论》	政治学

An Analysis of Odd Arne Westad's *The Global Cold War: Third World Interventions and the Making of Our Times*	解析文安立《全球冷战：美苏对第三世界的干涉与当代世界的形成》	政治学
An Analysis of Sigmund Freud's *The Interpretation of Dreams*	解析西格蒙德·弗洛伊德《梦的解析》	心理学
An Analysis of William James' *The Principles of Psychology*	解析威廉·詹姆斯《心理学原理》	心理学
An Analysis of Philip Zimbardo's *The Lucifer Effect*	解析菲利普·津巴多《路西法效应》	心理学
An Analysis of Leon Festinger's *A Theory of Cognitive Dissonance*	解析利昂·费斯汀格《认知失调论》	心理学
An Analysis of Richard H. Thaler & Cass R. Sunstein's *Nudge: Improving Decisions about Health, Wealth, and Happiness*	解析理查德·H.泰勒／卡斯·R.桑斯坦《助推：如何做出有关健康、财富和幸福的更优决策》	心理学
An Analysis of Gordon Allport's *The Nature of Prejudice*	解析高尔登·奥尔波特《偏见的本质》	心理学
An Analysis of Steven Pinker's *The Better Angels of Our Nature: Why Violence Has Declined*	解析斯蒂芬·平克《人性中的善良天使：暴力为什么会减少》	心理学
An Analysis of Stanley Milgram's *Obedience to Authority*	解析斯坦利·米尔格拉姆《对权威的服从》	心理学
An Analysis of Betty Friedan's *The Feminine Mystique*	解析贝蒂·弗里丹《女性的奥秘》	心理学
An Analysis of David Riesman's *The Lonely Crowd: A Study of the Changing American Character*	解析大卫·理斯曼《孤独的人群：美国人社会性格演变之研究》	社会学
An Analysis of Franz Boas's *Race, Language and Culture*	解析弗朗兹·博厄斯《种族、语言与文化》	社会学
An Analysis of Pierre Bourdieu's *Outline of a Theory of Practice*	解析皮埃尔·布尔迪厄《实践理论大纲》	社会学
An Analysis of Max Weber's *The Protestant Ethic and the Spirit of Capitalism*	解析马克斯·韦伯《新教伦理与资本主义精神》	社会学
An Analysis of Jane Jacobs's *The Death and Life of Great American Cities*	解析简·雅各布斯《美国大城市的死与生》	社会学
An Analysis of C. Wright Mills's *The Sociological Imagination*	解析C.赖特·米尔斯《社会学的想象力》	社会学
An Analysis of Robert E. Lucas Jr.'s *Why Doesn't Capital Flow from Rich to Poor Countries?*	解析小罗伯特·E.卢卡斯《为何资本不从富国流向穷国？》	社会学

An Analysis of Émile Durkheim's *On Suicide*	解析埃米尔·迪尔凯姆《自杀论》	社会学
An Analysis of Eric Hoffer's *The True Believer: Thoughts on the Nature of Mass Movements*	解析埃里克·霍弗《狂热分子：群众运动圣经》	社会学
An Analysis of Jared M. Diamond's *Collapse: How Societies Choose to Fail or Survive*	解析贾雷德·M. 戴蒙德《大崩溃：社会如何选择兴亡》	社会学
An Analysis of Michel Foucault's *The History of Sexuality Vol. 1: The Will to Knowledge*	解析米歇尔·福柯《性史（第一卷）：求知意志》	社会学
An Analysis of Michel Foucault's *Discipline and Punish*	解析米歇尔·福柯《规训与惩罚》	社会学
An Analysis of Richard Dawkins's *The Selfish Gene*	解析理查德·道金斯《自私的基因》	社会学
An Analysis of Antonio Gramsci's *Prison Notebooks*	解析安东尼奥·葛兰西《狱中札记》	社会学
An Analysis of Augustine's *Confessions*	解析奥古斯丁《忏悔录》	神学
An Analysis of C. S. Lewis's *The Abolition of Man*	解析 C. S. 路易斯《人之废》	神学

图书在版编目（CIP）数据

解析世界环境与发展委员会《布伦特兰报告：我们共同的未来》：汉、英/克塞尼亚·吉拉西莫娃（Ksenia Gerasimova）著；李莹莹译. 一上海：上海外语教育出版社，2020
（世界思想宝库钥匙丛书）
ISBN 978-7-5446-6139-3

Ⅰ.①解… Ⅱ.①克… ②李… Ⅲ.①环境保护—可持续性发展—研究—世界—汉、英 Ⅳ.①X22

中国版本图书馆CIP数据核字（2020）第022637号

This Chinese-English bilingual edition of *An Analysis of World Commission on Environment and Development's* The Brundtland Report: Our Common Future is published by arrangement with Macat International Limited.
Licensed for sale throughout the world.

本书汉英双语版由Macat国际有限公司授权上海外语教育出版社有限公司出版。
供在全世界范围内发行、销售。

图字：09 - 2018 - 549

出版发行：**上海外语教育出版社**
（上海外国语大学内） 邮编：200083
电　　话：021-65425300（总机）
电子邮箱：bookinfo@sflep.com.cn
网　　址：http://www.sflep.com
责任编辑：王　璐

印　　刷：上海信老印刷厂
开　　本：890×1240　1/32　印张 5.5　字数 114千字
版　　次：2020 年 8 月第 1 版　2020 年 8 月第 1 次印刷
印　　数：2 100 册

书　　号：ISBN 978-7-5446-6139-3
定　　价：30.00 元
本版图书如有印装质量问题，可向本社调换
质量服务热线：4008-213-263　电子邮箱：editorial@sflep.com